90.00

D1435506

This book is due for return on or before the last date shown below.

Atlas of the Mammalian Ovary

Katharina Spanel-Borowski

Atlas of the Mammalian Ovary

Morphological Dynamics and Potential Role of Innate Immunity

 Springer

Katharina Spanel-Borowski
Institute of Anatomy
University of Leipzig
Leipzig
Germany

ISBN 978-3-642-30534-4   ISBN 978-3-642-30535-1   (eBook)
DOI 10.1007/978-3-642-30535-1
Springer Heidelberg New York Dordrecht London

Library of Congress Control Number: 2012948617

Printed on acid-free paper

Springer is part of Springer Science+Business Media (www.springer.com)

Preface

The ovary is an organ with high tissue plasticity because of the growth, rupture, and atresia of follicles and the cyclic formation and decay of the corpus luteum (CL). The impressive dynamics in the ovary are especially evident in species of young reproductive age, but may be beyond the knowledge of pathologists who are familiar primarily with ovarian tumors and postmenopausal ovaries. Ovaries from small rodents, rabbits, dogs, and cows have estrous cycles of varying length and each of them offers specific advantages in drawing attention to particular ovarian events such as endocrine tumor development after whole-body X-irradiation and intraovarian oocyte release by superovulation.

This atlas on the mammalian ovary presents unique snapshots that reveal the complexity of follicular atresia, ovulation, as well as of CL formation and regression by providing insights into inflammatory responses and angiogenesis that go far beyond textbook level. Novel ways of rapidly regulating tissue homeostasis/turnover are termed cellular and vascular stromatolysis for the interstitial cortical tissue and vascular luteolysis for the regressing CL. Innate immunity (INIM) has the physiological ability to recognize dangerous places within the ovary and to precisely regulate inflammatory/anti-inflammatory processes under the premise of maintaining its cyclic function. INIM function also seems to remain quiescent during tissue remodeling. Cytokeratin-positive cells, considered as a novel dendritic-like cell type, appear to be the cellular defense arm of INIM in follicles, in CLs, and in the microvascular bed. The potential role of INIM in different ovarian events is briefly discussed in 11 separate chapters. Each chapter consists of a textual exposition and figure plates with legends. The figures are carefully described and provide deep insights into the dynamic world of ovarian morphology for both the beginner and the expert.

The atlas comes in time with the renaissance of picture appreciation, necessary to understand organ processes from the molecule to the whole. Morphology remains of key importance for basic and clinical research into ovarian disorders such as anovulation, polycystic ovaries, premature ovarian failure, autoimmune oophoritis, and ovarian tumors. I hope that this atlas on morphological dynamics will draw strong attention to the exciting and challenging field of INIM function in the ovary and that the work becomes important for cell biology, reproductive endocrinology, gynecology, and pathology.

Acknowledgments

The atlas is dedicated to my graduate students and colleagues in fond memory of shared times. I embrace my husband, Dr. R. Spanel, for the never-ending support and for the gentle push to do the atlas. Sincere thanks go to Isabella Athanassiou, BA, MPhil, Heidelberg, Germany, for language editing and to Mr. StephenMuthuRaj JoeArun, Chennai, India, for professional editing.

Contents

Abbreviations

AVT	Arginine vasotocin
CD	Cluster of differentiation
CGRP	Calcitonin gene-related peptide
CK	Cytokeratin
CL	Corpus luteum
COCs	Cumulus–oocyte complexes
DHEA	Dehydroepiandrosterone
EMBs	Endoplasmic-reticulum-derived multilamellar bodies
FSH	Follicle-stimulating hormone
hCG	Human chorionic gonadotropin
H&E	Hematoxylin and eosin
HOPA	Hematoxylin-orange-g-phosphomolybdic acid-aniline
HRP	Horseradish peroxidase
HIF-1α	Hypoxia-inducible factor
IFN-β/γ	Interferon- β/γ
IL-1/1β	Interleukin-1/1β
Il-1/6	Interleukin-1/6
INIM	Innate immunity
IOR	Intraovarian oocyte release
IVF	In vitro fertilization
KIT	Tyrosine kinase growth receptor; CD117
LC3	Microtubule light chain 3
LH	Luteinizing hormone
LOX-1	Lectin-like oxidized low-density receptor 1
MHC II	Major histocompatibility complex class II
Myd 88	Myeloid differentiation factor 88
NGF	Neurite growth factor
nLDL	Normal low-density lipoprotein
NOD-like R	Nucleotide oligomerization domain receptors
oxLDL	Oxidized low-density lipoprotein
PMSG	Pregnant mare serum gonadotropin
PRRs	Pattern recognition receptors
RAS	Renin–angiotensin
rEGF	Recombinant epidermal growth factor
rhCG	Recombinant human chorionic gonadotropin
ROS	Reactive oxygen species
s.c.	Subcutaneous
SP	Substance P
TGF-β	Transforming growth factor β family

TLR   Toll-like receptor
TLR4   Toll-like receptor 4
TNF-α Tumor necrosis factor-α
TRIF   TIR domain-containing adaptor protein inducing IFN-β
VEGF   Vascular endothelial cell growth factor

Introduction

1.1 Perspectives from the Past to the Present

Since Karl Ernst von Baer observed the oocyte in antral follicles under the light microscope in Königsberg/Kaliningrad in 1826, the mammalian ovary has provided fundamental topics that keep in step with the contemporary scientific field. At the end of the nineteenth century, insights into the morphological structures of the ovary were obtained using novel histological techniques. The complex process of growth and maturation of the oocyte could be linked to different follicle stages. The immature oocyte in the diplotene stage is resident in primordial follicles, the growing oocyte populates preantral and antral follicles, and the mature oocyte is seen in the preovulatory follicle called the Graaf follicle in memory of Regnier de Graaf (1641–1673). While a doctor at Delft, de Graaf was the first to describe antral follicles ("kleine bollekens") in "female testicles" (Jay 2000). de Graaf was also a pioneer in noticing globular bodies in the rabbit ovary after coitus and relating them to the number of offspring. The "bollekens" became the corpus luteum (CL) shortly thereafter. The yellow body of temporary endocrine nature develops from a preovulatory follicle after expulsion of the oocyte into the fallopian tube (Niswender et al. 2000). The color comes from carotenoids, especially lutein, probably via nutritional uptake. The knowledge of ovarian histology paved the way for great progress in the upcoming field of reproductive endocrinology. In the 1920s, Allen and Doisy collected several thousand milliliters of follicular fluid from hog ovaries to purify estrogen in crystalline form (Stephens and Moley 2009). The so-called follicle factor became the first sex/steroid hormone. A CL factor, which was assumed to regulate pregnancy, turned out to be another sex hormone named progesterone (Niswender 2002). The pure sex steroids allowed biochemical assays to be developed for serum analysis with the aim of understanding the hormonal regulation of the ovarian cycle. An estrogen rise depends on the dominant follicle, which is selected from a cohort of antral follicles in the mid-follicular phase (Baerwald et al. 2012). High progesterone levels are assigned to the CL phase. The hypothalamus and the pituitary gland are recognized as superior managers of the ovulatory event through cyclic secretion of gonadotropin-releasing factor, of follicle-stimulating hormone, and of luteinizing hormone (FSH, LH) (Lunenfeld 2004). It was gradually revealed that the hormonal communication requires activation of specific receptors, which trigger a complex signaling cascade (Tata 2005). Pharmacological inhibition of the releasing factor pathway generated the "pill" for family planning, which led to an unexpected sociocultural change (Tyler May 2010). Presently, powerful intraovarian regulators—which include signaling pathways of the transforming growth factor β (TGF-β) family and of the renin-angiotensin system (RAS) as well as of forkhead transcription factors—are accepted as a final force for follicle rupture (Richards and Pangas 2010; Uhlenhaut and Treier 2011; Gonçalves et al. 2012). These regulators are involved in the production of inflammatory cytokines such as interleukin-1β (IL-1β), interleukin-6 (Il-6), tumor necrosis factor-α (TNF-α), and colony-stimulating factors, which all peak in the preovulatory follicle (Adashi 1990; Makinoda et al. 2008). The cytokines are produced by follicle cells and by granulocytes and macrophages, which densely populate the preovulatory follicle wall (Best et al. 1996; Brännström and Enskog 2002). The sudden and striking leukocyte influx is finally associated with extracellular matrix degradation and capillary sprouting from the thecal cell layer into the granulosa cell layer at the time of follicle rupture. Sprouting depends on a network of angiogenic factors among which the paramount role of the vascular endothelial cell growth factor (VEGF)-receptor system and of the RAS family is recognized (Kaczmarek et al. 2005; Gonçalves et al. 2012). Activation of the system appears to be downstream of signaling through the hypoxia-inducible factor α (Kim and Johnson 2009). Leukocyte recruitment and angiogenesis in addition to edema and connective tissue degradation, which all characterize the ovulatory event, compare with a physiological inflammation (Espey 1994). This innovative concept has in the meantime been accepted by the scientific community and was recently modified as a specific

K. Spanel-Borowski, *Atlas of the Mammalian Ovary*,
DOI 10.1007/978-3-642-30535-1_1, © Springer-Verlag Berlin Heidelberg 2012

immunoresponse (Espey 2006; Richards et al. 2008). It is likely influential for CL formation too. As speculated, endothelin-2 and angiopietin-2, being vessel stabilizing factors, could interact in the process (Klipper et al. 2010; Nishigaki et al. 2011). The final architecture of the luteal microvascular bed depends on endothelial progenitor cells, which are present in the preovulatory follicle (Spanel-Borowski et al. 2007; Merkwitz et al. 2010). Somatic progenitor cells are separate from germ cell progenitors. In early fetal life, the surface epithelium is colonized by primordial germ cells. There is ongoing debate on whether the zone of the surface epithelium contains stem-cell-like precursors of oocytes able to proliferate and mature under the appropriate microenvironment (Begum et al. 2008; Bukovsky 2011; Virant-Klun et al. 2011). The scenario is emerging that menopausal ovaries can be used to obtain oocytes, which are grown in culture and mature in the presence of effective cytokine cocktails. After fertilization, totipotent embryoblasts could become embryonic stem cells to be used beneficially in regenerative medicine. Refined techniques will provide organ-specific cells for novel treatments of degenerative diseases. Collectively, events that are highly important for contemporary research take place in the ovary, which has maintained its front-line position in the field for more than 100 years.

1.2   Plasticity and Homeostasis

The intricate morphology of the ovary is subjected to continuously changing structures. These are follicles, CLs, as well as the cortex, which consists of interstitial gland cells (Oktem and Oktay 2008). There are ongoing changes in mammals, from fetal life to menopause. These changes are related to the growth and atresia of follicles until puberty and thereafter to follicle maturation, its rupture and transformation into a CL passing through various stages of development, secretion, and regression (Devoto et al. 2009; Baerwald et al. 2012). Interstitial gland cells look like endocrine cells or like fibroblasts depending on the endocrine status (Guraya 1978). Reorganization of the ovarian compartments requires that blood vessels, lymph vessels, autonomic nerve fibers, and connective tissue adjust to different demands. Apart from the endometrium, the ovary is the only organ in the adult organism that undergoes extremely high tissue plasticity in response to changing endocrine and paracrine stimulations. Tissue turnover by formation and regression of structures is well balanced. The size of the ovary remains stable during reproductive life in spite of ongoing folliculogenesis and CL formation. Additionally, the antral follicle cohort and CLs are dismissed in due time to permit the input of new structures for the next ovarian cycle. The task of controlling organ size and ensuring functional stability is made more difficult by repeated tissue wounding during the

reproductive period. Wounding and healing of tissues relate to cyclic follicle ruptures and to CL formation. The endocrine system alone is not equipped with appropriate tools to manage these enormous tasks. There must be a big brother/sister as a hidden authority, and innate immunity (INIM) is appearing on the horizon as the assumed mighty force. It has the physiological ability to regulate inflammatory and anti-inflammatory events in the ovulatory process.

1.3   Present Concept of INIM and Significance for the Ovary

Progress in immunity research has been in favor of adapted immunity because the growing expertise in T cell subtypes and antibody production allows novel strategies to be used for the therapy of immunological diseases. INIM has been a neglected field of research and was considered to be unspecific for a long time, because the function of INIM has been related to unspecific antimicrobicidal factors that are secreted since birth. The scene has changed dramatically, however, thanks to the discovery of pattern recognition receptors (PRRs) (Turvey and Broide 2010). They comprise four families with the receptor names of toll-like (TLR), NOD-like, RIG-like, and C-lectin-like (O'Neill and Bowie 2007; Kumar et al. 2009). The TLR gene, first discovered in *Drosophila* as a gene for dorsoventral body orientation, is translated into a transmembrane protein with a cytoplasmic portion similar to the IL-1 receptor protein. Its activation triggers a signaling pathway with the ability to activate around 3,000 genes. The choice of cytoplasmic adaptor proteins decides on pathways mediated by the myeloid differentiation factor 88 (Myd88) or by the TIR domain-containing adaptor protein inducing IFN-β (TRIF) resulting in inflammation or anti-inflammation (Takeuchi and Akira 2010). The exoplasmic portion of PRRs recognizes evolutionarily conserved clusters of lipoproteins, lipopolysaccharides, peptidoglycans, nucleic acids, and mRNA from bacteria and viruses altogether termed pathogen-associated patterns. The PRRs also respond to danger-/damage-associated ligands called "alarmins" because of their origin from stressed and dying cells (Bianchi 2007; Rock et al. 2010). The danger ligands comprise acute-phase proteins such as heat shock proteins, S100 calcium-binding proteins, high mobility box group 1 protein, hyaluronan fragments, amyloid peptides, and uric acids. The discovery of PRRs runs in parallel to changes in the immunological concept of self and nonself detection (Matzinger 2002; Köhl 2006a, b). Nonself detection has been attributed to T cells recognizing pathogens as foreign invaders. Thereafter, the role of dendritic cells in support of T cell function has become evident (Bancereau and Steinman 1998; Mellman and Steinman 2001). Dendritic cells produce abundant PRRs and respond to pathogen-associated patterns and to danger ligands

(Hoshino and Kaisho 2008). Because dendritic cells belong to the cellular defense arm of INIM, the concept that INIM provides the first response against pathogens was put forward. In the next big step, pathogens from outside the body were put on a similar level with danger signals from inside the body. Any wounding, contusion, and surgical intervention with tissue disruption and bleeding releases danger signals/alarmins from dying and dead cells, and dendritic cells react by physiological inflammation (Matzinger 2002; Medzhitov 2008, 2010). The danger concept of immune defense explains the evolution of the immune system as continuous adaptations against dangerous events. Innate immunity is more than 500 million years old and still the only defense system in worms (Endo et al. 2006; Turvey and Broide 2010). The adaptive immune system of mammals came about many thousands of years later as a younger brother. Sensing danger from inside or outside the body thus awakens INIM function, whereas absence of exogenous and endogenous danger signals does not do so. Any inflammation and tumor growth is tolerated by INIM as long as it feels there is no challenge for the organism (Medzhitov 2010). Dangerous processes might heal through a controlled activation of the INIM-dependent signaling cascade (Chan and Housseau 2008). Overactivation will cause allergic diseases, autoimmune diseases, and transplant rejection (Steinman and Banchereau 2007).

Each organ seems to shape the immunological tolerance according to its own needs (Niederkorn 2006; Matzinger 2007; Matzinger and Kamala 2011). This implies that immunological training has individually taken place during organ development. The antigen-presenting dendritic cells as a major arm of the INIM executive appear to have the capacity to supply a versatile organ-specific immune protection. Dendritic cells are heterogeneous according to anatomical location, mission, and origin (Banchereau and Steinman 1998; Mellman and Steinman 2001). Dendritic cells are recruited as immature cells from their places of origin into the blood, migrate to the danger/antigen-presenting site, and then mature on their way toward the regional lymph node. Here, processing of antigens triggers two pathways of PRRs: one for gene activation and the other for antigen presentation together with the major histocompatibility complex (MHC) (Takeda and Akira 2005). The cell surface complex of dendritic cells becomes the docking station for naïve T cells, which subsequently differentiate into cytotoxic and helper cells. Dendritic cells thus bridge INIM with adaptive immunity function (Iwasaki and Medzhitov 2010; Schenten and Medzhitov 2011). Classic dendritic cells are differentiated into lymphoid and nonlymphoid lineage with origin from lymphocyte or plasmacytoid precursor cells of the bone marrow. The dorsal yolk sac region also comes into focus as a novel dendritic cell source, because the region hosts a microglia cell subtype with dendritic-like function according to tracing studies in early embryogenesis of mice (Ginhoux

et al. 2010). The dorsal yolk sac is close to the aortolumbar region and the genital ridge, which appears to give rise to another dendritic-like cell characterized by a transient expression of cytokeratin (CK). The CK-positive cells arise from the fetal surface epithelium and the sex cords cells, both being densely populated by germ cells (Löffler et al. 2000; Spanel-Borowski 2011a). In childhood the CK-positive granulosa cells disappear during folliculogenesis, and they reappear in the preovulatory follicle and in the bovine CL during development. Human CK-positive follicle cells regulate TLR4 under in vitro stimulation with oxidized low-density lipoprotein (oxLDL) (Serke et al. 2009, 2010). The CK-positive cells isolated from the bovine CL remain intact under interferon-γ (IFN-γ) treatment. Molecules of adhesion plates and of tight junctions are up-regulated as is the major histocompatibility complex class II (MHC II) (Spanel-Borowski and Bein 1993; Fenyves et al. 1993, 1994; Ricken et al. 1996). Because of the analogy to IFN-γ-treated dendritic cells (Billiau and Matthys 2009), CK-positive cells are judged as dendritic-like cells with guardian function in the mammalian ovary. These potential dendritic cells could regulate and convey specific INIM function in the ovary.

1.4 Ovarian Events Under Potential INIM Control

Growth and atresia of follicles, follicle rupture, and the CL life cycle correlate with remodeling events of different intensities. Cell decay and tissue degradation might represent sites of varying dangers. They may or may not cause inflammatory-like immunoreactions as an expression of INIM function (Medzhitov 2008, 2010). In the following section, ovarian events are looked at under the aspect of potential INIM control.

Primordial follicles, primary follicles, and small preantral follicles reside in the human ovary at birth. In childhood, small antral follicles grow under the control of intraovarian factors and sex hormones (Craig et al. 2007). From puberty to menopause, a cohort of large antral follicles arise at the beginning of each ovarian cycle, and the preovulatory follicle is selected in the middle follicular phase to become the preovulatory follicle (Fig. 1.1) (Baerwald et al. 2012). Recruitment of antral follicles and selection of the dominant follicle and its maturation demands stimulation by FSH and LH to initiate effective intraovarian pathways (Craig et al. 2007; Richards and Pangas 2010; Rodgers and Irving-Rodgers 2010). Thus, the gonadotropin-independent follicle growth in the back should be distinguished from a gonadotropin-dependent event in the front. It is associated with a well-developed microvascular bed in the thecal cell layer for supply of gonadotropins (Fraser and Duncan 2005; Kaczmarek et al. 2005). The density and architecture of the microvessels probably control which one out of the cohort of antral follicles will be selected to become the definitive winner. Follicles grow as perfect spheres. It signifies a tightly controlled kinetic program in the follicle wall to add a comparable amount of cells at the poles and between the poles. The follicle wall should be similar in thickness to guarantee that biomolecules reach the oocyte from all directions in similar concentrations and that oocyte-derived regulating factors of the transforming growth factor-β family (TGF-β) influence the follicle microenvironment uniformly (Juengel et al. 2006a; Conti et al. 2011). The three-dimensional coordination of follicle growth as a sphere is not understood at all, although folliculogenesis under the influence of sex steroids and growth factors has been extensively studied (Findlay et al. 2001; Juengel et al. 2006b; Pangas 2007). A small preantral follicle with low estrogen production resides in the periphery of the cortex. Follicle growth coincides with increasing concentrations of estrogens in the follicular fluid and in the serum. "Ingrowth" from the periphery toward the medullary part of the cortex occurs. The interstitial cortical tissue itself appears to tolerate the expansion process and might even support expansion through activation of matrix metalloproteinases (Robinson et al. 2001; Goldman and Shalev 2004). The remodeling process during folliculogenesis is obviously recognized as a nondangerous process by INIM sensors.

Oocyte atresia begins in primordial follicles of the human fetus at gestation month 6. The term becomes "follicular atresia" when preantral and antral follicles develop after birth. Cessation of oogenesis at gestation month 7 and the continuous process of atresia are responsible that out of seven million fetal oocytes only 400 become fertilizable in a woman considering a 40-year period of reproductive life. Apoptosis of granulosa and thecal cells through the Fas-Fas ligand signaling system appears to be a major molecular mechanism causing antral follicle death (Matsuda-Minehata et al. 2006; Craig et al. 2007). Necrosis and cell-death autophagy are emerging as additional cell death forms in this follicle stage (van Wezel et al. 1999; Rodgers and Irving-Rodgers 2010). Ignorance still exists about the molecular pathway of atresia in preantral follicles. They maintain intact follicle cells, while the oocyte shows eosinophilic necrosis with zona pellucida rupture (Spanel-Borowski 1981). The preantral oocyte obviously shows an increased susceptibility to an altered microenvironment compared to the antral oocyte, which acquires competence in a complex series of coordinated steps (Telfer and McLaughlin 2007).

The ovulatory period extends in women over roughly 12 h with the rupture time point in the middle. Detachment of the oocyte with cumulus expansion and resumption of first meiosis is known to occur in the preovulatory follicle. Ovulation relates to the 1-h-long follicle rupture with extrusion of the oocyte into the fallopian tube. In the postovulatory phase, the rupture site of the follicle wall is closed by a fibrin clot, thrombi occur in the adjacent microvessels, capillaries sprout from the theca into the avascular granulosa cell

layer, and leukocytes are recruited (Espey 1994). The ovulatory changes are accepted as physiological inflammation with precisely controlled steps of connective tissue degradation and of healing. The search to understand the driving force of the acute event is turning its attention toward INIM surveillance as a physiological immunoresponse to danger (Richards et al. 2002, 2008; Spanel-Borowski 2011a, b). There is evidence that the preovulatory follicle is a structure under oxidative stress and that ovulation represents a danger-signaling event of INIM interference. It is finally initiated by high amounts of reactive oxygen species (ROS) being byproducts of LH-dependent full-speed steroidogenesis. ROS comes from the mitochondrial respiratory chain and is released through leaky membranes (Hanukoglu 2006). In a vicious cycle, ROS causes oxidation of lipoprotein to become oxLDL, which binds to the lectin-like oxLDL receptor 1 (LOX-1) of granulosa cells. The subsequent signaling cascade, which is associated with augmented LOX-1 and ROS production (Mehta et al. 2006; Chen et al. 2007), causes apoptotic death in CK-negative granulosa cells (Serke et al. 2009, 2010). Nonapoptotic cell death is also initiated by oxLDL after binding to TLR4, which is expressed by CK-positive granulosa cells. The 20–50 % of dead granulosa cells in fresh follicle harvests of women under in vitro fertilization (IVF) therapy reflects the danger of oxidative stress (Vilser et al. 2010).

The CL develops from the ruptured follicle. Rupture is followed by complete degradation of the basement membrane. Granulosa cells and thecal cells become large and small steroidogenic lutein cells, respectively, being associated with sprouting capillaries (Niswender et al. 2000; Reynolds et al. 2000). The fully developed CL consists of roughly 50 % of vascular cells. Functional luteolysis happens around day 24 of the ovarian cycle in women and immediately terminates progesterone secretion without histological changes in the CL (Devoto et al. 2009). Structural luteolysis follows and capillaries are the first to die at the onset of regression. The CL stages of formation, secretion, and regression depend on tight molecular control. Stages are terminated with great precision in time to allow the income of the next follicle cohort and the next CL cycle (Stocco et al. 2007). Because of the availability of bovine CLs from slaughterhouses and because of its similarity to the human CL, the bovine CL is a beautiful model for gaining insights into angiogenesis and angioregression, into the growth factor profile, and into how the profile changes in time (Schams and Berisha 2004; Kaczmarek et al. 2005). Cell death pathways of luteolysis are insufficiently explained by the apoptotic mechanism alone. Recent findings point to the additional role of autophagy through which damaged organelles and apoptotic bodies are removed by autodigestion (Gaytán et al. 2008; Choi et al. 2011). The disintegration process—which lasts for several ovarian cycles in, for example, rats, cows, and women—compares with chronic inflammation because of heavy macrophage infiltration, maintenance of large-sized microvessels, and hyalinization of connective tissue (Bauer et al. 2003; van Wu et al. 2004). A mighty coordinator is needed, who is able to build a CL and to dismiss it by controlled angioregression and cell death in due time. The big boss seems to be an INIM force that coordinates the development and maintenance of the CL. The endocrine system is considered as an excellent coplayer. This novel concept has been recently put forward (Spanel-Borowski 2011a, c). Structural luteolysis starts when dendritic-like cells present alarmins from hypoxia-damaged luteal cells to naïve T cells and thus call in the function of adaptive immunity. Altogether, INIM control without or with bridging to adaptive immunity arises as a mighty force in the ovary. To date, INIM has been hidden behind the endocrine system. Both of these powerful systems interact with each other in a totally unknown network that guarantees precise repetitions of ovarian cycles during the reproductive period.

1.5 Material, Methods, and Aims

Ovaries were derived from species with a short estrous cycle such as, mice, rats, and golden hamsters (Spanel-Borowski 2011a). A long estrous/ovarian cycle occurs in dogs, cows, and humans. Rabbits were used as induced stimulators. Each species provided different advantages for varying approaches in this study. For example, intravenous pulse labeling with 3^H thymidine was used to assess cell proliferation in the dog ovary (Spanel-Borowski et al. 1984). Whole-body X-irradiation was feasible in dogs for studying the recuperation time in the reappearance of preovulatory follicles (Spanel-Borowski and Calvo 1982). Intravenous injection of polyester resin could be performed in golden hamsters to obtain vessel corrosion casts of the ovaries (Spanel-Borowski et al. 1987). Immature hamsters and rats were superovulated to synchronize the estrous cycle for the study of intraovarian oocyte release (IOR) in serial sections (Spanel-Borowski et al. 1982, 1983; Löseke and Spanel-Borowski 1996). Preovulatory follicles and CLs from cows were available from the local slaughterhouse for immunostaining of serial sections and for cell cultures (Spanel-Borowski and van der Bosch 1990; Spanel-Borowski et al. 1997; Spanel-Borowski 2011b). Cells were harvested from follicle aspirates of women under IVF therapy and paraffin blocks of human ovaries were collected from archives of the Institutes of Pathology at Ulm, Heidelberg, and Leipzig (Heider et al. 2001; Vilser et al. 2010).

The methods and experiments for cell proliferation and whole-body X-irradiation in dogs (Spanel-Borowski et al. 1981; Spanel-Borowski and Calvo 1982), for superovulation in mice, rats, and rabbits (Spanel-Borowski et al. 1986), and for dehydroepiandrosterone (DHEA) treatment to obtain polycystic ovaries in rats (Krishna et al. 2001) have been previously described and are cited in the figure legends. In brief, ovaries

were embedded in paraffin wax and serially sectioned. For 3^H thymidine labeling, autoradiography was conducted with serial sections. In general, staining was performed with hematoxylin and eosin (H&E), for azan, and with the tetra-chrome staining by hematoxylin–orange-g–phosphomolybdic acid–aniline (HOPA). Indirect immunostaining or immunofluorescence localization for diverse antigens was individually performed with paraffin sections, cryostat sections, and cell cultures derived from the bovine CL. The classic technique of ultrastructure was extended to ultrahistochemistry for detection of horse radish peroxidase (HRP) activity in sprouting capillaries (Spanel-Borowski and Mayerhofer 1987). The gold labeling technique was used for albumin localization in rat follicles (Krishna and Spanel-Borowski 1989). Microvessel corrosion casts were investigated under the scanning electron microscope (Spanel-Borowski et al. 1987).

The art of taking pictures in histology and in electron microscopy has been a powerful instrument in the last century. The technique is now neglected because molecular insights have opened up novel fields and they are making great progress in extending the knowledge of organ function and in developing novel therapy strategies. Nevertheless, the renaissance of picture appreciation is a result of the necessity to understand development, from the single molecule to the whole organism. Our morphological observations on ovarian dynamics were made over about 30 years by studying around 100,000 serial sections. Details of three-dimensional changes are captured by thoroughly examining serial sections. The time will never come back for such a silent type of work. Although the growth and atresia of follicles, ovulation, the CL life cycle with stages of formation, secretion, and regression are well noted in reproductive biology, special features of these major events are widely unknown. The pathologist is familiar with the simple structure of the postmenopausal ovary, because "young" ovaries are seldom obtained from young women by surgical intervention or after a fatal accident. Therefore, morphological dynamics pass unnoticed in ovaries of young reproductive age. It appears helpful to fill the gap by observations from "young" ovaries of different mammalian species. Any disturbance of the complex dynamics in diverse ovarian events might finally contribute to the high variety of tumor types in the ovary (Tavassoli and Devilee 2003) signaling an altered INIM function. In the postmenopausal ovary, INIM function could fade and generate conditions for tumor growth. This atlas presents images of remodeling events as possible footmarks of INIM function and of places of danger. Footmarks can be related to different cell death forms, to the presence of eosinophils and mast cells, to inflammatory-like patterns with fibrin thrombi, sprouting, and regressing capillaries, and to the existence of CK-positive cells as a potential novel type of dendritic cell. The major aim of this book is to make clear that the ovary is anything but a quiescent organ. The complexity of morphological dynamics and tissue plasticity in the ovary is unique for an adult organism.

1.6 Scheme for Follicle Growth and Corpus Luteum Life Cycle

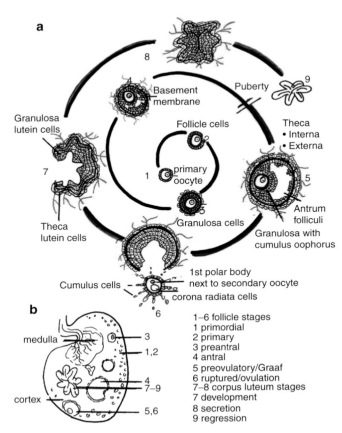

Fig. 1.1 Stages of follicle growth and maturation as well as of the CL life cycle are drawn in a spiral-like manner and as overview (**a** and **b**). (**a**) Basal follicle growth is the background event from birth to puberty. The zona pellucida starts to develop in the primary follicle, and becomes distinct as a homogeneous layer between granulosa cells and the primary oocyte in the small preantral follicle. The thecal cell layer of the larger preantral follicle is vascularized and separated from the avascular granulosa by a basement membrane. In the reproductive period, the ovarian cycle with the follicle phase, the ovulatory phase, and the different CL stages develops regularly. The ovulatory phase culminates in the expulsion of the secondary oocyte through the ruptured follicle wall undergoing luteinization and angiogenesis. (**b**) The cortex is populated by follicles and by CLs. Immature/young follicles reside in the periphery below the surface epithelium. Follicle growth is associated with expansion into the medullary part of the cortex. The medulla contains the supplying vessels (Drawn by Christine Feja, Leipzig)

References

Adashi EY (1990) The potential relevance of cytokines to ovarian physiology: the emerging role of resident ovarian cells of the white blood cell series. Endocr Rev 11:454–464

Baerwald A, Adams G, Pierson R (2012) Ovarian antral folliculogenesis during the human menstrual cycle: a review. Hum Reprod Update 18:73–91

Banchereau J, Steinman RM (1998) Dendritic cells and the control of immunity. Nature 392:245–252

Bauer M, Schilling N, Spanel-Borowski K (2003) Development and regression of non-capillary vessels in the bovine corpus luteum. Cell Tissue Res 311:199–205

Begum S, Papaioannou VE, Gosden RG (2008) The oocyte population is not renewed in transplanted or irradiated adult ovaries. Hum Reprod 23:2326–2330

Best CL, Pudney J, Welch WR, Burger N, Hill JA (1996) Localization and characterization of white blood cell populations within the menstrual cycle and menopause. Hum Reprod 11:790–797

Bianchi ME (2007) DAMPs, PAMPs and alarmins: all we need to know about danger. J Leukoc Biol 81:1–5

Billiau A, Matthys P (2009) Interferon-gamma: a historical perspective. Cytokine Growth Factor Rev 20:97–113

Brännström M, Enskog A (2002) Leukocyte networks and ovulation. J Reprod Immunol 57:47–60

Bukovsky A (2011) Ovarian stem cell niche and follicular renewal in mammals. Anat Rec (Hoboken) 294:1284–1306

Chan CW, Housseau F (2008) The 'kiss of death' by dendritic cells to cancer cells. Cell Death Differ 15:58–69

Chen CJ, Kono H, Golenbock D, Reed G, Akira S, Rock KL (2007) Identification of a key pathway required for the sterile inflammatory response triggered by dying cells. Nat Med 13:851–856

Choi J, Jo M, Lee E, Choi D (2011) The role of autophagy in corpus luteum regression in the rat. Biol Reprod 85:465–472

Conti M, Hsieh M, Musa Z, Oh JS (2011) Novel signaling mechanisms in the ovary during oocyte maturation and ovulation. Mol Cell Endocrinol 356:65–73

Craig J, Orisaka M, Wang H, Orisaka S, Thompson W, Zhu C, Kotsuji F, Tsang BK (2007) Gonadotropin and intra-ovarian signals regulating follicle development and atresia: the delicate balance between life and death. Front Biosci 12:3628–3639

Devoto L, Fuentes A, Kohen P, Cespedes P, Palomino A, Pommer R, Munoz A, Strauss JF 3rd (2009) The human corpus luteum: life cycle and function in natural cycles. Fertil Steril 92:1067–1079

Endo Y, Takahashi M, Fujita T (2006) Lectin complement system and pattern recognition. Immunobiology 211:283–293

Espey LL (1994) Current status of the hypothesis that mammalian ovulation is comparable to an inflammatory reaction. Biol Reprod 50:233–238

Espey LL (2006) Comprehensive analysis of ovarian gene expression during ovulation using differential display. Methods Mol Biol 317:219–241

Fenyves AM, Behrens J, Spanel-Borowski K (1993) Cultured microvascular endothelial cells (MVEC) differ in cytoskeleton, expression of cadherins and fibronectin matrix. A study under the influence of interferon-gamma. J Cell Sci 106(Pt 3):879–890

Fenyves AM, Saxer M, Spanel-Borowski K (1994) Bovine microvascular endothelial cells of separate morphology differ in growth and response to the action of interferon-gamma. Experientia 50:99–104

Findlay J, Britt K, Kerr JB, O'Donnell L, Jones ME, Drummond AE, Simpson ER (2001) The road to ovulation: the role of oestrogens. Reprod Fertil Dev 13:543–547

Fraser HM, Duncan WC (2005) Vascular morphogenesis in the primate ovary. Angiogenesis 8:101–116

Gaytán M, Morales C, Sanchez-Criado J, Gaytán F (2008) Immunolocalization of beclin 1, a bcl-2-binding, autophagy-related protein, in the human ovary: possible relation to life span of corpus luteum. Cell Tissue Res 331:509–517

Ginhoux F, Greter M, Leboeuf M, Nandi S, See P, Gokhan S, Mehler MF, Conway SJ, Ng LG, Stanley ER, Samokhvalov I, Merad M (2010) Fate mapping analysis reveals that adult microglia derive from primitive macrophages. Science 330:841–845

Goldman S, Shalev E (2004) MMPS and TIMPS in ovarian physiology and pathophysiology. Front Biosci 9:2474–2483

Gonçalves P, Ferreira R, Gasperin B, Oliveira J (2012) Role of angiotensin in ovarian follicular development and ovulation in mammals: a review of recent advances. Reproduction 143:11–20

Guraya SS (1978) Recent advances in the morphology, histochemistry, biochemistry, and physiology of interstitial gland cells of mammalian ovary. Int Rev Cytol 55:171–245

Hanukoglu I (2006) Antioxidant protective mechanisms against reactive oxygen species (ROS) generated by mitochondrial P450 systems in steroidogenic cells. Drug Metab Rev 38:171–196

Heider U, Pedal I, Spanel-Borowski K (2001) Increase in nerve fibers and loss of mast cells in polycystic and postmenopausal ovaries. Fertil Steril 75:1141–1147

Hoshino K, Kaisho T (2008) Nucleic acid sensing toll-like receptors in dendritic cells. Curr Opin Immunol 20:408–413

Iwasaki A, Medzhitov R (2010) Regulation of adaptive immunity by the innate immune system. Science 327:291–295

Jay V (2000) A portrait in history. The legacy of Reinier de Graaf. Arch Pathol Lab Med 124:1115–1116

Juengel J, Reader K, Bibby A, Lun S, Ross I, Haydon L, McNatty K (2006a) The role of bone morphogenetic proteins 2, 4, 6 and 7 during ovarian follicular development in sheep: contrast to rat. Reproduction 131:501–513

Juengel J, Heath D, Quirke L, McNatty K (2006b) Oestrogen receptor alpha and beta, androgen receptor and progesterone receptor mRNA and protein localisation within the developing ovary and in small growing follicles of sheep. Reproduction 131:81–92

Kaczmarek M, Schams D, Ziecik A (2005) Role of vascular endothelial growth factor in ovarian physiology – an overview. Reprod Biol 5:111–136

Kim J, Johnson R (2009) You don't need a PHD to grow a tumor. Dev Cell 16:781–782

Klipper E, Levit A, Mastich Y, Berisha B, Schams D, Meidan R (2010) Induction of endothelin-2 expression by luteinizing hormone and hypoxia: possible role in bovine corpus luteum formation. Endocrinology 151:1914–1922

Köhl J (2006a) Self, non-self, and danger: a complementary view. Adv Exp Med Biol 586:71–94

Köhl J (2006b) The role of complement in danger sensing and transmission. Immunol Res 34:157–176

Krishna A, Spanel-Borowski K (1989) Intracellular detection of albumin in the ovaries of golden hamsters by light and electron microscopy. Arch Histol Cytol 52:387–393

Krishna A, al Rifai A, Hubner B, Rother P, Spanel-Borowski K (2001) Increase in calcitonin gene related peptide (CGRP) and decrease in mast cells in dihydroepiandrosterone (DHEA)-induced polycystic rat ovaries. Anat Embryol (Berl) 203:375–382

Kumar H, Kawai T, Akira S (2009) Toll-like receptors and innate immunity. Biochem Biophys Res Commun 388:621–625

Löffler S, Horn LC, Weber W, Spanel-Borowski K (2000) The transient disappearance of cytokeratin in human fetal and adult ovaries. Anat Embryol (Berl) 201:207–215

Löseke A, Spanel-Borowski K (1996) Simple or repeated induction of superovulation: a study on ovulation rates and microvessel corrosion casts in ovaries of golden hamsters. Ann Anat 178:5–14

Lunenfeld B (2004) Historical perspectives in gonadotrophin therapy. Hum Reprod Update 10:453–467

Makinoda S, Hirosaki N, Waseda T, Tomizawa H, Fujii R (2008) Granulocyte colony-stimulating factor (G-CSF) in the mechanism of human ovulation and its clinical usefulness. Curr Med Chem 15:604–613

Matsuda-Minehata F, Inoue N, Goto Y, Manabe N (2006) The regulation of ovarian granulosa cell death by pro- and anti-apoptotic. J Reprod Dev 52:695–705

Matzinger P (2002) The danger model: a renewed sense of self. Science 296:301–305

Matzinger P (2007) Friendly and dangerous signals: is the tissue in control? Nat Immunol 8:11–13

Matzinger P, Kamala T (2011) Tissue-based class control: the other side of tolerance. Nat Rev Immunol 11:221–230

Medzhitov R (2008) Origin and physiological roles of inflammation. Nature 454:428–435

Medzhitov R (2010) Inflammation 2010: new adventures of an old flame. Cell 140:771–776

Mehta JL, Chen J, Hermonat PL, Romeo F, Novelli G (2006) Lectin-like, oxidized low-density lipoprotein receptor-1 (LOX-1): a critical player in the development of atherosclerosis and related disorders. Cardiovasc Res 69:36–45

Mellman I, Steinman RM (2001) Dendritic cells: specialized and regulated antigen processing machines. Cell 106:255–258

Merkwitz C, Ricken AM, Lösche A, Sakurai M, Spanel-Borowski K (2010) Progenitor cells harvested from bovine follicles become endothelial cells. Differentiation 79:203–210

Niederkorn JY (2006) See no evil, hear no evil, do no evil: the lessons of immune privilege. Nat Immunol 7:354–359

Nishigaki A, Okada H, Tsuzuki T, Cho H, Yasuda K, Kanzaki H (2011) Angiopoietin 1 and angiopoietin 2 in follicular fluid of women undergoing a long protocol. Fertil Steril 96:1378–1383

Niswender GD, Juengel JL, Silva PJ, Rollyson M, McIntush EW (2000) Mechanisms controlling the function and life span of the corpus luteum. Physiol Rev 80:1–29

Niswender G (2002) Molecular control of luteal secretion of progesterone. Reproduction 123:333–339

Oktem O, Oktay K (2008) The ovary: anatomy and function throughout human life. Ann N Y Acad Sci 1127:1–9

O'Neill LA, Bowie AG (2007) The family of five: TIR-domain-containing adaptors in toll-like receptor. Nat Rev Immunol 7:353–364

Pangas S (2007) Growth factors in ovarian development. Semin Reprod Med 25:225–234

Reynolds LP, Grazul-Bilska AT, Redmer DA (2000) Angiogenesis in the corpus luteum. Endocrine 12:1–9

Richards JS, Russell DL, Ochsner S, Espey LL (2002) Ovulation: new dimensions and new regulators of the inflammatory-like response. Annu Rev Physiol 64:69–92

Richards JS, Liu Z, Shimada M (2008) Immune-like mechanisms in ovulation. Trends Endocrinol Metab 19:191–196

Richards J, Pangas S (2010) The ovary: basic biology and clinical implications. J Clin Invest 120:963–972

Ricken A, Rahner C, Landmann L, Spanel-Borowski S (1996) Bovine endothelial-like cells increase intercellular junctions under treatment with interferon-gamma. An in vitro study. Ann Anat 178:321–330

Robinson LL, Sznajder NA, Riley SC, Anderson RA (2001) Matrix metalloproteinases and tissue inhibitors of metalloproteinases in human fetal testis and ovary. Mol Hum Reprod 7:641–648

Rock KL, Latz E, Ontiveros F, Kono H (2010) The sterile inflammatory response. Annu Rev Immunol 28:321–342

Rodgers RJ, Irving-Rodgers HF (2010) Morphological classification of bovine ovarian follicles. Reproduction 139:309–318

Schams D, Berisha B (2004) Regulation of corpus luteum function in cattle–an overview. Reprod Domest Anim 39:241–251

Schenten D, Medzhitov R (2011) The control of adaptive immune responses by the innate immune system. Adv Immunol 109: 87–124

Serke H, Vilser C, Nowicki M, Hmeidan FA, Blumenauer V, Hummitzsch K, Lösche A, Spanel-Borowski K (2009) Granulosa cell subtypes respond by autophagy or cell death to oxLDL-dependent activation of the oxidized lipoprotein receptor 1 and toll-like 4 receptor. Autophagy 5:991–1003

Serke H, Bausenwein J, Hirrlinger J, Nowicki M, Vilser C, Jogschies P, Hmeidan F, Blumenauer V, Spanel-Borowski K (2010) Granulosa cell subtypes vary in response to oxidized low-density lipoprotein as regards specific lipoprotein receptors and antioxidant enzyme activity. J Clin Endocrinol Metab 95:3480–3490

Spanel-Borowski K (1981) Morphological investigations on follicular atresia in canine ovaries. Cell Tissue Res 214:155–168

Spanel-Borowski K (2011a) Footmarks of innate immunity in the ovary and cytokeratin-positive cells as potential dendritic cells, vol 209, Advances in anatomy, embryology and cell biology. Springer, Heidelberg. ISBN 978-3-642-16076-9

Spanel-Borowski K (2011b) Ovulation as danger signaling event of innate immunity. Mol Cell Endocrinol 333:1–7

Spanel-Borowski K (2011c) Five different phenotypes of endothelial cell cultures from the bovine corpus luteum: present outcome and role of potential dendritic cells in luteolysis. Mol Cell Endocrinol 338:38–45

Spanel-Borowski K, Calvo W (1982) Short- and long-term response of the adult dog ovary after 1200 R whole-body X-irradiation and transfusion of mononuclear leukocytes. Int J Radiat Biol Relat Stud Phys Chem Med 41:657–670

Spanel-Borowski K, Mayerhofer A (1987) Formation and regression of capillary sprouts in corpora lutea of immature superstimulated golden hamsters. Acta Anat (Basel) 128:227–235

Spanel-Borowski K, van der Bosch J (1990) Different phenotypes of cultured microvessel endothelial cells obtained from bovine corpus luteum. Study by light microscopy and by scanning electron microscopy (SEM). Cell Tissue Res 261:35–47

Spanel-Borowski K, Bein G (1993) Different microvascular endothelial cell phenotypes exhibit different class I and II antigens under interferon-gamma. In Vitro Cell Dev Biol Anim 29A:601–602

Spanel-Borowski K, Trepel F, Schick P, Pilgrim C (1981) Aspects of cellular proliferation during follicular atresia in the dog ovary. Cell Tissue Res 219:173–183

Spanel-Borowski K, Petterborg LJ, Reiter RJ (1982) Preantral intraovarian oocyte release in the white-footed mouse, *Peromyscus leucopus*. Cell Tissue Res 226:461–464

Spanel-Borowski K, Vaughan LY, Johnson LY, Reiter RJ (1983) Increase of intra-ovarian oocyte release in PMSG-primed immature rats and its inhibition by arginine vasotocin. Biomed Res 4:71–82

Spanel-Borowski K, Thor-Wiedemann S, Pilgrim C (1984) Cell proliferation in the dog (beagle) ovary during proestrus and early estrus. Acta Anat (Basel) 118:153–158

Spanel-Borowski K, Sohn G, Schlegel W (1986) Effects of locally applied enzyme inhibitors of the arachidonic acid cascade on follicle growth and intra-ovarian oocyte release in hyperstimulated rabbits. Arch Histol Jpn 49:565–577

Spanel-Borowski K, Amselgruber W, Sinowatz F (1987) Capillary sprouts in ovaries of immature superstimulated golden hamsters: a SEM study of microcorrosion casts. Anat Embryol (Berl) 176:387–391

Spanel-Borowski K, Rahner P, Ricken AM (1997) Immunolocalization of CD18-positive cells in the bovine ovary. J Reprod Fertil 111:197–205

Spanel-Borowski K, Sass K, Loffler S, Brylla E, Sakurai M, Ricken AM (2007) KIT receptor-positive cells in the bovine corpus luteum are primarily theca-derived small luteal cells. Reproduction 134:625–634

Steinman RM, Banchereau J (2007) Taking dendritic cells into medicine. Nature 449:419–426

Stephens S, Moley K (2009) Follicular origins of modern reproductive endocrinology. Am J Physiol Endocrinol Metab 297:E1235–E1236

Stocco C, Telleria C, Gibori G (2007) The molecular control of corpus luteum formation, function, and regression. Endocr Rev 28:117–149

Takeda K, Akira S (2005) Toll-like receptors in innate immunity. Int Immunol 17:1–14

Takeuchi O, Akira S (2010) Pattern recognition receptors and inflammation. Cell 140:805–820

Tata J (2005) One hundred years of hormones. EMBO Rep 6:490–496

Tavassoli FA, Devilee P (2003) Pathology and genetics of tumours of the breast and female genital organs. International Agency for Research on Cancer, Lyon

Telfer E, McLaughlin M (2007) Natural history of the mammalian oocyte. Reprod Biomed Online 15:288–295

Turvey SE, Broide DH (2010) Innate immunity. J Allergy Clin Immunol 125:S24–S32

Tyler May E (2010) America+the pill. A history of promise, peril and literature. Basic Books, New York, NY

Uhlenhaut N, Treier M (2011) Forkhead transcription factors in ovarian function. Reproduction 142:489–495

Van Wezel IL, Dharmarajan AM, Lavranos TC, Rodgers RJ (1999) Evidence for alternative pathways of granulosa cell death in healthy and slightly atretic bovine aantral follicles. Endocrinology 140:2602–2612

Van Wu R, der Hoek KH, Ryan NK, Norman RJ, Robker RL (2004) Macrophage contributions to ovarian function. Hum Reprod Update 10:119–133

Vilser C, Hueller H, Nowicki M, Hmeidan FA, Blumenauer V, Spanel-Borowski K (2010) The variable expression of lectin-like oxidized low-density lipoprotein receptor (LOX-1) and signs of autophagy and apoptosis in freshly harvested human granulosa cells depend on gonadotropin dose, age, and body weight. Fertil Steril 93:2706–2715

Virant-Klun I, Stimpfel M, Skutella T (2011) Ovarian pluripotent/multipotent stem cells and in vitro oogenesis in mammals. Histol Histopathol 26:1071–1082

Primordial follicles host the immature/primary oocyte, which is surrounded by a flat discontinuous monolayer of epithelial cells (Rodgers and Irving-Rodgers 2010). In the human, primordial follicles appear with the onset of meiosis in the fourth month of pregnancy, and the oocyte is arrested in the diplotene stage around birth (Baerwald et al. 2012). The pool of primordial follicles resides in the periphery of the ovarian cortex. The subsequent stages are associated with ingrowth of the developing follicles into the medullary section of the cortex (Fig. 2.1a, b). The international nomenclature of mammalian follicles prefers "preantral follicles" instead of secondary follicles. They comprise small, medium, and large follicles with average diameters between 50 and 250 µm in small rodents, dogs, cows, and humans. The increase in follicle size is associated with oocyte growth from roughly 40 to 110 µm in diameter and with the change from a monolayer of cuboidal follicle cells into a multilayer of granulosa cells (Fig. 2.1c–e). The name of the latter comes from the granula-like size of the nurse cells in comparison to the big oocyte, which belongs to the largest cells of the body apart from the motor neuron. Antral follicles represent the subsequent follicle stages with an average diameter of between 251 and 600 µm. The term "antral follicle" refers to the fluid-filled antrum more precisely than "tertiary follicle." Antral follicles are recognized by a distinct external and internal thecal cell layer and a well-developed microvascular bed (Fig. 2.1f–h). The development of preantral follicles starts shortly after birth, and antral follicles are still found in a 3-year-old child. Basic follicle growth exclusively depends on intraovarian growth factors such as the TGF-β family and the RAS system (Pangas 2007; Richards and Pangas 2010) as well as on moderate estrogen concentrations for local needs (Findlay et al. 2001; Juengel et al. 2006a). Any gonadotropic influence is missing until puberty. At that time, rising concentrations of FSH and the pulsatile release of LH from the pituitary gland trigger the first ovarian cycle (Craig et al. 2007). The preovulatory follicle, also termed Graaf follicle, is selected from a cohort of large antral follicles. High levels of estrogen are responsible for many proliferating cells in the outer granulosa cell layer of preovulatory follicles compared with preantral and antral follicles. In Fig. 2.2, proliferating cells are detected in the dog ovary after intravenous ^{3}H thymidine labeling and autoradiography (Spanel-Borowski et al. 1984).

A perfect follicle requires coordinated granulosa cell proliferation to become an impeccable sphere with the oocyte as center. Granulosa cell layers of comparable number at the poles and in between the poles guarantee that biomolecules reach the oocyte in similar concentrations (Juengel et al. 2006b; Conti et al. 2011). The relationship between oocyte size and follicle layers can be disturbed. Irregular interactions between the two partners seem to result in inadequate follicle growth. Multi- and unilamellar layers of granulosa cells at opposite poles generate an elliptic form of the oocyte with the larger diameter being adjusted to flat poles (Fig. 2.3a, b). Uneven poles and deformation of the oocyte can be associated with a striking transformation of cuboidal cells into columnar granulosa cells similar to the "mucification" of corona radiata cells in the cumulus expansion of a preovulatory follicle (Fig. 2.3c, d). Premature antrum formation by apoptosis of granulosa cells is noted in medium-sized preantral follicles with an oocyte of antral follicle size and with two deformed oocytes, respectively (Fig. 2.3e, f). Absence of complete antrum formation in multilamellar follicles of antral follicle size and with a prominent thecal cell layer might be caused by an insufficient oocyte being too small and of eccentric position (Fig. 2.3g, h).

Inadequate follicle growth reflects inadequate cell kinetics. It is not yet considered as a troublemaker responsible for the follicle pool reduction. Follicular atresia as the other side of the coin is held responsible for the decrease in follicles starting as oocyte atresia around gestation month 6 in the human (Oktem and Oktay 2008). Out of seven million oocytes in fetal life, 99.9 % are doomed to die by oocyte/follicular atresia until menopause. For pathway A, regressing preantral follicles maintain granulosa cells, while the oocyte becomes damaged (Spanel-Borowski 1981). It is likely that rupture of the zona pellucida harms the ooplasm. It becomes irregular in shape, showing eosinophilia and fragmentation as signs of nonapoptotic/necrotic cell death (Fig. 2.4). Intact granulosa cells invade the ooplasm, which gradually disappears leaving remnants of the zona pellucida in an interstitial cell cord (Fig. 2.5a–d). Prominent remodeling of the columnar cells can reshape the cell cord into a pseudotubular structure of atrophic stage. A hyalinized zona pellucida in the cortical stroma represents the terminal stage of a regressed preantral follicle (Fig. 2.5e–f). For pathway B, regressing antral follicles follow a different program. They have acquired an oocyte and a zona pellucida of enormous robustness, while granulosa cells undergo apoptosis (Fig. 2.6a–d). When apoptotic bodies have disappeared most likely by phagocytosis, an uneven granulosa cell layer of changing thickness indicates the process of regression. Apoptosis of thecal cells is less evident, but could explain zones without separation between internal and external thecal cell layers. They can undergo different degrees of hypertrophy. The follicle cyst represents an advanced stage of antral follicle atresia. One discontinuous layer of flat or cuboidal granulosa cells lines the antrum hosting an intact looking oocyte. When a "naked/free oocyte" with an intact zona pellucida appears in the cortical stroma, the structure gives testimony of previous antral follicle atresia (Fig. 2.6e, f).

The two completely different atretic pathways A and B signify a different molecular pattern for survival of the oocyte and granulosa cells, respectively. Pathway A appears to be without danger for the granulosa cell community. Pathway B, however, provides the molecular program of apoptosis in follicle cells by using the Fas-Fas ligand pathway (Matsuda-Minehata et al. 2006; Hussein 2005; Inoue et al. 2011). It is speculated that the development of PRRs coincides with antral follicle stages and that interactions with estrogen receptors occur. The estrogen receptors are widely studied in different follicle stages and obviously interfere with the apoptotic pathway (Findlay et al. 2001; Juengel et al. 2006a; Liew et al. 2011). In view of the opposite behavior of oocytes in pathways A and B, it appears that the zona pellucida protects the oocyte. When cross-linking of the zona pellucida matrix is still fragile in preantral follicles, rupture occurs. The oocyte loses protection through a closed microenvironment and necrotic cell death is initiated. The stability of the zona pellucida in antral follicles is attributed to a firm matrix and to its mechanical protection through the follicular fluid. Thus, the zona pellucida is stable when the oocyte grows from the preantral to the antral follicle stage. This point has not yet been studied at the molecular level, while data on the necrotic or apoptotic pathway of cell death are available for the oocyte (Perez et al. 2007; Kujjo et al. 2011). Antral follicles under atresia pathway B are rescued by gonadotropic stimulation of the ovary, and the related follicle aspirates contain healthy looking oocytes. The outcome of an IVF procedure could differ for oocytes harvested from intact or from rescued follicles.

Radiation effects in Beagle dog ovaries confirm and extend the observations on follicular atresia by inadequate growth and by pathways A and B (Fig. 2.7). Adult dogs survive lethal 1,200-R (12 Gy) whole-body X-irradiation because of mononuclear cell transfusion at the time of intervention (Spanel-Borowski and Calvo 1982). Short-term radiolesions up to 3 weeks after the exposure show columnar metaplasia of granulosa cells as premature mucification in preantral and antral follicles (Fig. 2.7a, b). Pathway B occurs in preantral follicles showing apoptotic granulosa cells and an intact zona pellucida–oocyte unit. The irregular occurrence of apoptotic granulosa cells explains the cell poles as well as their circumscribed loss resulting in the oocyte coming into direct contact with the cortical stroma (Fig. 2.7c–f). Proliferating granulosa cells are sensitive to radiation, because it represents a DNA-damaging agent and induces apoptotic cell death (Roos and Kaina 2012). It is possible that irradiation causes up-regulation of the *Bax* gene in preantral follicles. An increase of the *Bax* gene and of apoptosis was reported for preantral granulosa cells, which were obtained from mice after treatment with 4-vinylcyclohexene diepoxide (Tilly 1996; Springer et al. 1996). Alternatively, the *insulin-like factor* gene known to decrease the apoptotic cascade in mice follicles (Spanel-Borowski et al. 2001) could be deleted by whole-body X-irradiation. Radiation appears not to affect the oocytes of preantral follicles, which is difficult to understand. Radiation may stabilize the molecular architecture of the zona pellucida in such way that the oocyte remains protected against environmental stress. In line with this thought is the morphology of the zona pellucida, which looks thicker after irradiation than without it (compare Figs. 2.1 and 2.7). Pathway B in preantral follicles is associated with transformation of follicle cells into cells of the interstitial cortical stroma (Fig. 2.7d). It signifies a friendly interaction between the regressing follicle and the stroma in spite of danger signals from dying granulosa cells. The good tissue compliance of the two different cell types could explain the presence of preantral follicles with two oocytes (Fig. 2.7g, h). The phenomenon, which is recurring, is likely due to partial fusion of adjacent follicles.

Long-term radiolesions are classified into effects I and II. Radiation effect I, seen in Beagle dogs up to 73 days after irradiation, results in severely damaged antral follicles (Fig. 2.8). They show a modified atresia pathway B after production of a heavily hyalinized basement membrane also called "glassy membrane" (Spanel-Borowski and Calvo 1982). Its presence is attributed to the excessive accumulation of ground components released from many dying follicle cells. Terminal stages of pathway B relate to round oocytes surrounded by a thick zona pellucida and a gyrated glassy membrane (Fig. 2.8d–f). Radiation effect I displays a cortical stroma with interstitial gland cells of fibroblast-like, inactive appearance (Chap. 4). In contrast, in effect II, which is detected up to 217 days after irradiation, interstitial gland cells have changed into the steroidogenic-secreting cell type (Fig. 2.9a; Chap. 4). This might be caused by irradiation-dependent activation of the neurosecretory process in the hypothalamus (Duchesne et al. 1968). Granulosa cell cords, representative of atresia pathway A in preantral follicles, accumulate in density and lead to columnar metaplasia of granulosa cells (Fig. 2.9a, b). It appears that tumors such as Leydig-like cell tumor, Sertoli–Leydig-like cell tumor, and Sertoli-like cell tumor arise from granulosa cell cords around 7 months after whole-body X-irradiation (Fig. 2.9c–e). The granulosa cell-like tumor could originate from previous primary and primordial follicles, which lose the oocyte but maintain prominent follicle cells (Fig. 2.9a, f). Ovarian tumors belong to the category of sex cord tumors according to the classification of the World Health Organization (Tavassoli and Devilee 2003). Our observations thus point to the possibility that disturbed follicular atresia can result in endocrine ovarian tumors. The risk of ovarian tumor development increases after menopause when ovarian cycles cease (Kurman and Shih 2010). One neglected cause could be that INIM function declines in the ovary.

The dog ovary response to the lethal whole-body X-irradiation dose remains surprising. In spite of severe damage of all follicle types, some primordial follicles start to grow 60 days after irradiation and pass through various follicle stages until becoming mature follicles (Spanel-Borowski and Calvo 1982). The first ovulation is seen about 170 days later. Corpora lutea are found and pregnancies occur in subsequent years. Resumption of reproductive capability is also reported for species whose ovaries depict a high radioresistance to a sublethal dose of whole-body X-irradiation

(Erickson et al. 1976). The resumption of the ovarian cycle could be attributed to the rescue of some primordial follicles (Spanel-Borowski and Calvo 1982). This assumption is in conflict with findings that the primary follicle reserve is completely ablated after exposure to high-dose whole-body X-irradiation and that neo-folliculogenesis fails in mice after sterilization by chemicals or by γ-irradiation (Kerr et al. 2012). Beagle dogs survive the lethal irradiation dose, because mononuclear cells are transfused at the time of intervention. The procedure to purify monocytes might have also isolated epiblast-like stem cells, which could have been mobilized into the peripheral blood during blood collection (Ratajczak et al. 2011). These particular cells could play an important role as a back-up population of tissue-committed stem cells by migrating into the zone of the irradiation-deleted follicle reserve. The epiblast-like cells, which are assumed to be deposited at the onset of gastrulation in developing tissues (Ratajczak et al. 2011), could have the potential to proliferate and mature into oocytes in the appropriate microenvironment (Begum et al. 2008; Bukovsky 2011; Virant-Klun et al. 2011). Development of the postulated oocyte stem cells might require the ovarian INIM network, which also has to recuperate from irradiation effects. At present this hypothesis will be perceived as a scientific impossibility. In the very long run, however, this highly speculative concept could develop into an innovative strategy to maintain fertility in young women who suffer from a malignant tumor and face chemotherapy. Today, cryopreservation of the ovary and transplantation is one of the therapy options available (Silber 2012).

This chapter ends with a scheme on variations of atresia pathways A and B (Fig. 2.10).

2.1   Follicles as a Compartment of Cell Proliferation

2.1.1   Intact Follicles

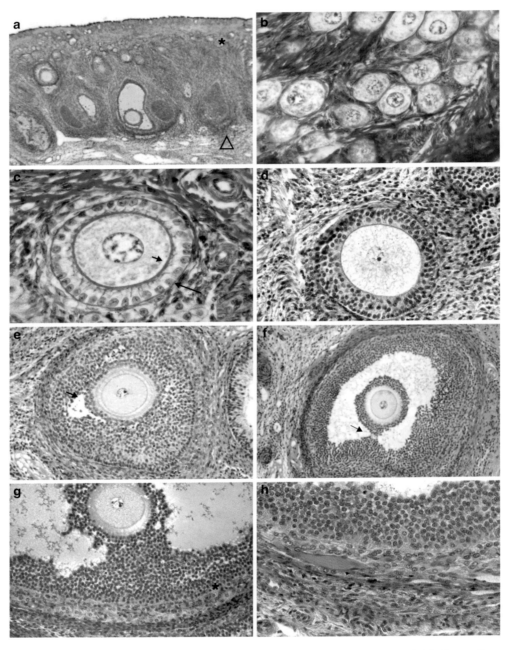

Fig. 2.1 Growth stages from primordial to antral follicles are shown for the mature beagle ovary. (**a**) The overview of the cortex displays the surface epithelium with tunica albuginea above nests of primordial follicles (*asterisk*). Stages of preantral and antral follicles develop from the intermediary toward the medullary part of the cortex. The medulla contains blood and lymph vessels in addition to nerve fibers in a loose fibrocollagenous tissue (*arrowhead*). (**b**) Primordial follicles with squamous follicular cells populate the closely packed, spindle-shaped fibroblast-like cells of the cortex. The primary oocytes are arrested in the diplotene stage of first meiotic division. (**c**) The follicle with cuboidal follicular cells is called the unilaminar primary follicle. The amorphous zona pellucida separates the oocyte from the follicular cells (*short arrow*). Stromal cells begin to organize as thecal cells close to the basement membrane (*long arrow*). (**d**) The small preantral follicle represents a multilamellar secondary follicle of more than three layers of granulosa cells. (**e**) The large preantral follicle with more than ten layers and with a fluid-filled cavity (*arrow*) marks the transition from the secondary/preantral to the tertiary/antral follicle. (**f**) The granulosa of the medium-sized antral follicle subdivides into the mural/outer part with projection of the cumulus oophorus (*arrow*) into the antrum. The inner part contains the corona radiata cells that are immediately adjacent to the oocyte (**g, h**). The thecal cell layer of the antral follicle depicts the theca interna with prominent microvessels and steroid-producing cells (*asterisk*), whereas the theca externa is mostly composed of fibroblasts and myofibroblasts. H&E (**a, d–h**); Azan (**b, c**), × a:30; b,c:280; d,h:180; e,g.120; f:80 (Adapted from Spanel-Borowski 2010)

2.1.2   **Cell Proliferation**

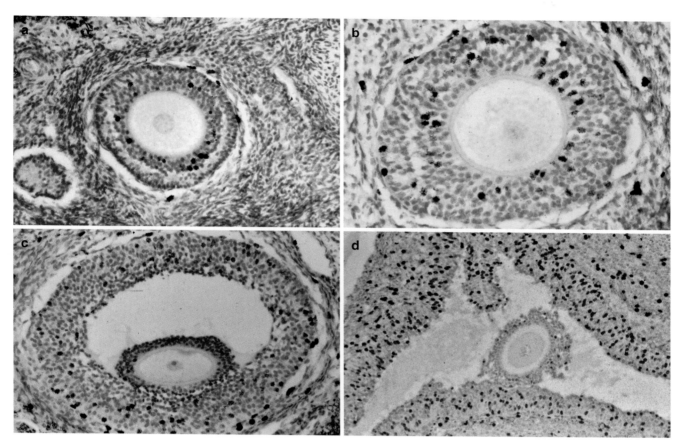

Fig. 2.2 Cell proliferation is restricted to follicles as noted in autoradiographs after ³H-thymidine pulse labeling of mature beagles (0.6–1.5 mCi/kg body weight). (**a–c**) The percentage of labeled granulosa cells (*black grains* on nuclei) is below 5 % in intact preantral and antral follicles. Cell proliferation in the cortical stroma is negligible in (**a**). (**d**) The labeling index increases to 15 % in the outer/mural granulosa of a preovulatory follicle. In comparison, the inner granulosa shows a decrease in the number of labeled cells. x a,c:150;b:230;d:80 (Adapted from Spanel-Borowski et al. 1984)

2.1.3 Inadequate Growth

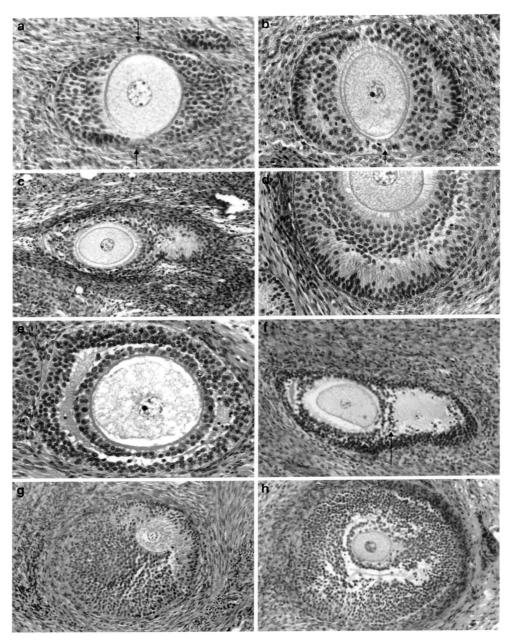

Fig. 2.3 Signs of inadequate follicle growth are depicted for the mature beagle ovary. (**a**, **b**) The granulosa of the preantral follicles shows irregular thickness of the opposite poles. The single- or two-layer poles of the granulosa coincide with the oocyte's larger diameter (*arrows*). Pole formation and elliptic oocyte form impede correct follicle growth as a sphere. (**c**, **d**) Pole formation in the preantral follicles is associated with premature mucification. The outer granulosa layer transforms into columnar epithelial cells (similar to mucification of corona radiata cells in a preovulatory follicle; Fig. 5.9d). (**e**, **f**) Premature antrum formation (*asterisk*) in preantral follicles occurs in the presence of a spherical oocyte in (**e**) or two deformed oocytes in (**f**) Apoptotic granulosa cells (*arrow*) are untypical for this follicle stage (Fig. 2.6). (**g**, **h**) The oocytes of two large preantral follicles are deformed and too small in diameter for the follicle stage. The eccentric oocyte position is accompanied by premature mucification in (**g**) or premature antrum formation in (**h**). H&E, x a,b,d,e:180;c,f,g,h:110

2.2 Follicular Atresia

2.2.1 Preantral Follicles with Pathway A

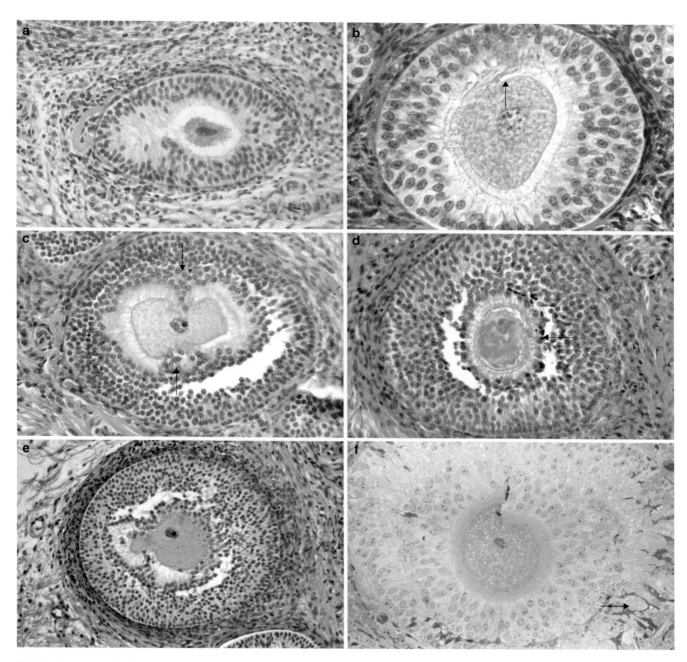

Fig. 2.4 Preantral follicles regress by atresia pathway A with nonapoptotic death of the oocyte as displayed for the mature beagle ovary. (**a**) The oocyte of the preantral follicle is shrunk, the nucleus is condensed, and the ooplasm is eosinophilic. (**b**) Rupture of the zona pellucida with extrusion of ooplasm (*arrow*). (**c**) Invasion of granulosa cells into the ooplasm through the zona pellucida, which has ruptured twice (*arrows*). The cavity represents a fixation/embedding artifact. (**d**) The dying oocyte with eosinophilic plaques and the rupture of the zona pellucida are shown. (**e**) Remnants of the zona pellucida adjust at the knob-like extension of the heavily damaged oocyte. (**f**) Rupture of the zona pellucida is accompanied by the appearance of dark cells of dendritic-like shape (*arrow*) in the Richardson-stained semithin section (Fig. 3.5a, b). H&E (**a–e**), x a,c,d.230; b:350, e:150;f:280 (Adapted from Spanel-Borowski 1981)

2.2.2 Preantral Follicles with Advanced Pathway A

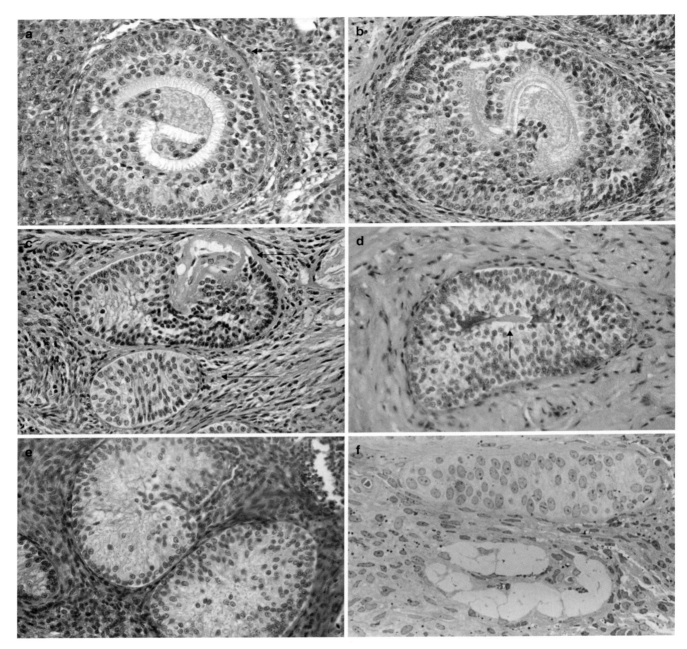

Fig. 2.5 The progression from final to terminal stages of atresia pathway A is depicted for preantral follicles in the mature beagle ovary. (**a**, **b**) Invasion of granulosa cells and extrusion of ooplasm have deformed the ruptured zona pellucida, which resembles typefaces. The basement membrane begins to hyalinize in (**a**) (*arrow*). (**c**) The collapsed zona pellucida is eliminated through the basement membrane (*short arrow*). The final outcome is indicated by the granulosa cell cord (*long arrow*). (**d**) The granulosa cell cord contains residues of the zona pellucida (*arrow*). (**e**) Pseudotesticular tubules, which look atrophic, arise by metaplasia into single- or two-layered columnar epithelial cells. (**f**) A granulosa cell cord is seen adjacent to a collapsed zona pellucida depicting the former rupture site in the Richardson-stained section. H&E (**a–e**), x a,b,c,d,e.230; f.370

2.2.3 Antral Follicles with Pathway B

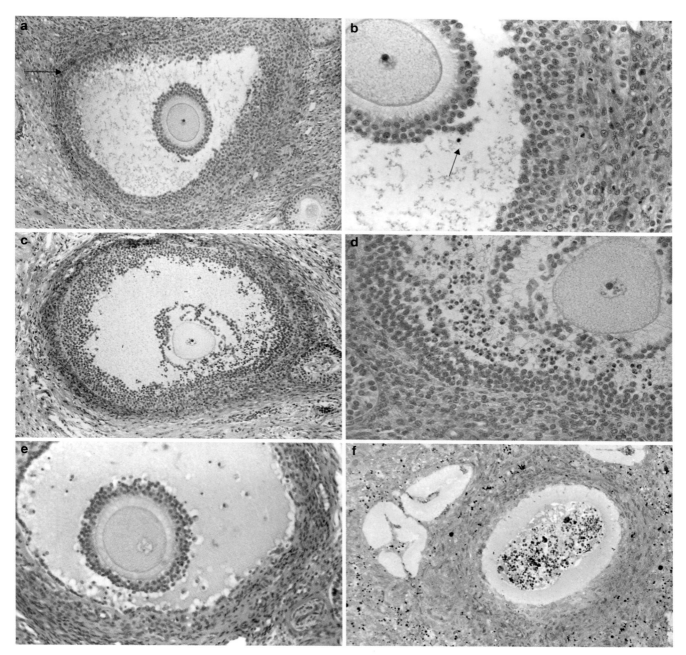

Fig. 2.6 Antral follicles with atresia pathway B in particular undergo apoptotic death of granulosa cells, as derived from the mature beagle ovary. (**a, b**) At the very early stage of atresia pathway B, the mural granulosa layer is irregular in thickness. Loosening of the granulosa cell layer occurs together with restricted disappearance of the theca interna and externa (**a**, *arrow*). An apoptotic body is noted close to inner granulosa cells (**b**, *arrow*). The healthy looking oocyte shows an eccentric nucleolus in the nucleus. (**c, d**) At the advanced stage of atresia pathway B, many apoptotic bodies are associated with thinning of the inner and outer granulosa cell layers. Degeneration of the thecal cell layer has progressed leading to more than half of the follicle wall without subdivision into theca interna and externa. The oocyte is slightly deformed (**d**). (**e**) The follicle cyst represents the final stage of pathway B. The mural granulosa is reduced to a cuboidal/squamous monolayer, whereas inner granulosa cells and the oocyte appear intact. (**f**) The "free/naked" oocyte shows an intact zona pellucida and accumulated lipid droplets in the Richardson-stained semithin section. The free oocyte indicates the terminal stage of atresia pathway B after resorption of the follicular fluid and transformation of intact follicle cells into the cortical stroma (Fig. 3.2a–c). Two collapsed and ruptured zonae pellucidae mark terminal stages of atresia pathway A in preantral follicles. H&E (**a–e**), x a,c:80; b,d:230; e:150, f:370 (Adapted from Spanel-Borowski 1981)

2.2.4   1,200-R Whole-body X-irradiation

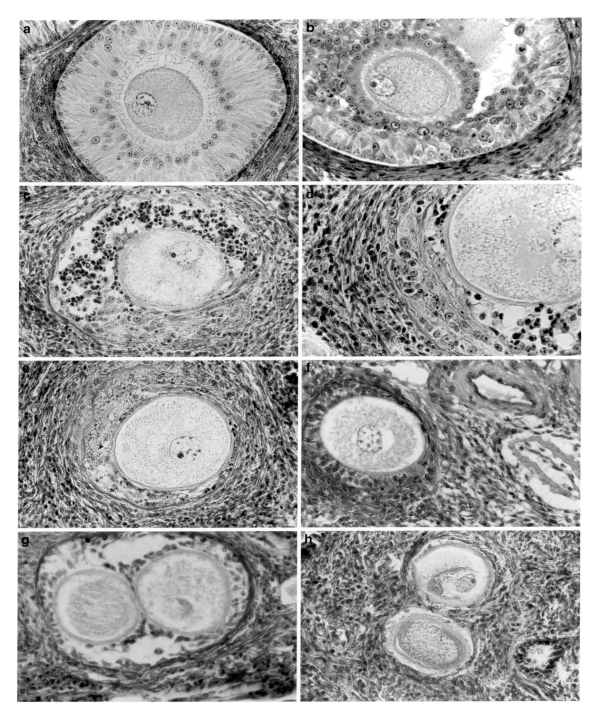

Fig. 2.7 1,200-R (12 Gy) whole-body X-irradiation of adult beagles together with transfusion of mononuclear cells alters follicles on day 10 after the exposure. Immediate radiolesions relate to inadequate follicle growth and to atresia pathway B in preantral and antral follicles. (**a, b**) High columnar epithelial cells indicate metaplasia and premature mucification in the preantral follicle and the antral one. The changes occur preferentially in the outer granulosa cells compared to the oocyte-associated cells. (**c, d**) Apoptotic granulosa cells in the presence of a healthy looking oocyte surrounded by a thick zona pellucida are untypical for preantral follicles. Still-intact granulosa cells transform into interstitial gland cells in (**d**). (**e, f**) Inadequate growth of a preantral follicle because of pole formation could be caused by circumscribed granulosa cell death. A few apoptotic cells are noted in (**e**). (**g, h**) Fusion of two adjacent preantral follicles might have led to the follicle with two oocytes in either the early or the advanced stage of atresia. The basement membrane disappears on contact with oocytes. HOPA, x a,b,c,e,f,g,h:200; d:310 (Adapted from Spanel-Borowski and Calvo 1982)

2.2.5 Effect I After 1,200-R Whole-body X-irradiation

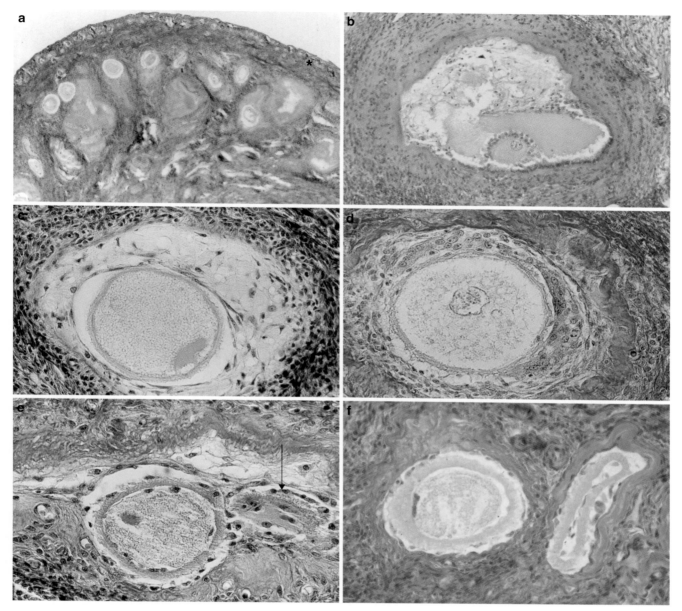

Fig. 2.8 1,200-R (12 Gy) whole-body X-irradiation of adult beagles together with transfusion of mononuclear cells generates radiation effect I on day 10 after the exposure. (**a**) A small cortical rim of inactive/degenerated interstitial cortical tissue is maintained below the zone of primordial and primary follicles (*asterisk*). Remnants of antral follicles appear with heavily hyalinized basement membranes. (**b–d**) Antral follicles in advanced stages of atresia either with an antrum in (**b**), with connective tissue replacement or residues of granulosa cells in the former antrum in (**c**) and (**d**). A multinuclear cell is noted in the former granulosa in (**e**) (*arrow*). The "glassy membrane" represents the hyalinized basement membrane (**d–f**). The "glassy membranes" become gyrated after antrum collapse and surround a "naked" oocyte with a thick zona pellucida in the terminal stages of atresia in (**f**). HOPA, x a:30; b.80; c,d,e,f.370 (Adapted from Spanel-Borowski and Calvo 1982)

2.2.6   Effect II After 1,200-R Whole-body X-irradiation

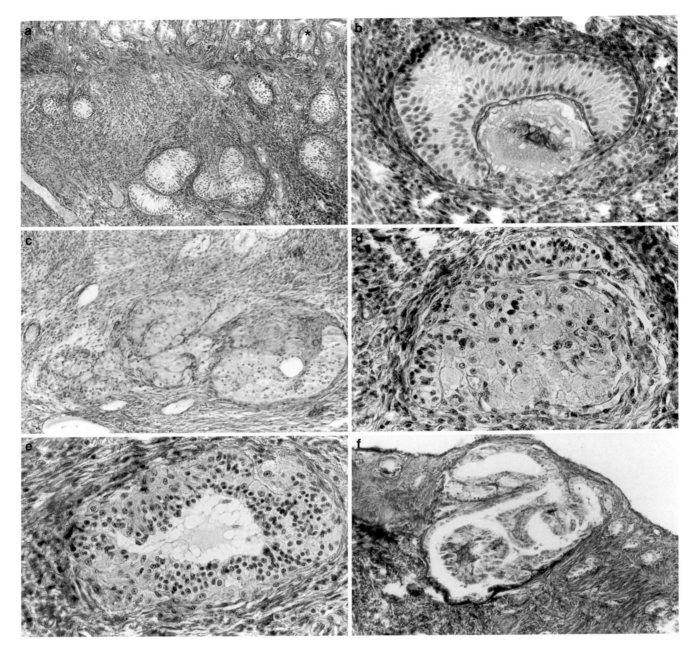

Fig. 2.9 1,200-R (12 Gy) whole-body X-irradiation of adult beagles together with transfusion of mononuclear cells causes radiation effect II on day 10 after the exposure. Granulosa cell cords undergo hypertrophy and transform into pseudotubules in the atrophic state. Thus, pathway A of follicular atresia has occurred. Tumors develop from regressing follicles 195 days after irradiation in (**c–f**). (**a**) The cortical zone of the previous primordial and primary follicles depicts crowded cuboidal epithelial cells (*asterisk*). Oocytes are not apparent. The cortex is dominated by mature interstitial gland cells and contains granulosa cell cords undergoing columnar metaplasia. (**b**) A pseudotesticular tubule embraces a thick zona pellucida. (**c–e**) Tumors, which are likely derived from granulosa cell cords, are reminiscent of a well-differentiated Leydig cell tumor in (**c**), a Sertoli–Leydig cell tumor in (**d**), and a Sertoli cell tumor in (**f**). (**e**) The granulosa cell tumor might have originated from the columnar epithelium derived from previous primordial/primary follicles. HOPA, x a,f:30; b,d,e:240; c:80 (Adapted from Spanel-Borowski and Calvo 1982)

2.3   Scheme of Pathways A and B

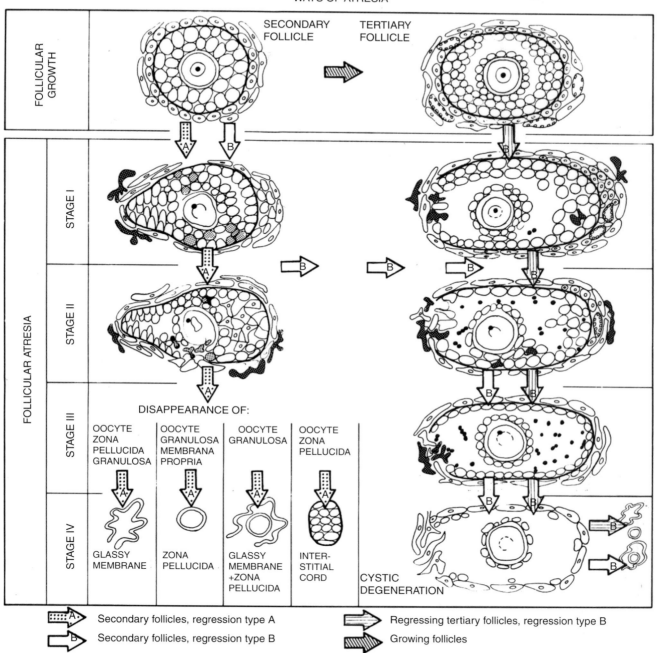

Fig. 2.10 The scheme depicts atresia pathways A and B in preantral and antral follicles, respectively. Findings from cyclic beagle ovaries and from ovaries after 1,200-R /12 Gy) whole-body X-irradiation are summarized for the terminal stages (stage IV). The membrana propria represents the basement membrane. Premature granulosa cell apoptosis in preantral follicles switches pathway A to pathway B

References

Baerwald A, Adams G, Pierson R (2012) Ovarian antral folliculogenesis during the human menstrual cycle: a review. Hum Reprod Update 18:73–91

Begum S, Papaioannou V, Gosden RG (2008) The oocyte population is not renewed in transplanted or irradiated adult ovaries. Hum Reprod 23:2326–2330

Bukovsky A (2011) Ovarian stem cell niche and follicular renewal in mammals. Anat Rec (Hoboken) 294:1284–1306

Conti M, Hsieh M, Musa Z, Oh JS (2011) Novel signaling mechanisms in the ovary during oocyte maturation and ovulation. Mol Cell Endocrinol 356:65–73

Craig J, Orisaka M, Wang H, Orisaka S, Thompson W, Zhu C, Kotsuji F, Tsang BK (2007) Gonadotropin and intra-ovarian signals regulating follicle development and atresia: the delicate balance between life and death. Front Biosci 12:3628–3639

Duchesne PY, Hajdukovic S, Beaumariage ML, Bacq ZM (1968) Neurosecretion in the hypothalamus and posterior pituitary after irradiation and injection of chemical radioprotectors in the rat. Radiat Res 34:583–595

Erickson BH, Reynolds RA, Murphree RL (1976) Late effects of 60Co gamma radiation on the bovine oocyte as reflected by oocyte survival, follicular development, and reproductive performance. Radiat Res 68:132–137

Findlay JK, Britt K, Kerr JB, O'Donnell L, Jones ME, Drummond AE, Simpson ER (2001) The road to ovulation: the role of oestrogens. Reprod Fertil Dev 13:543–547

Hussein MR (2005) Apoptosis in the ovary: molecular mechanisms. Hum Reprod Update 11:162–177

Inoue N, Matsuda F, Goto Y, Manabe N (2011) Role of cell-death ligand-receptor system of granulosa cells in selective follicular atresia in porcine ovary. J Reprod Dev 57:169–175

Juengel J, Heath D, Quirke L, McNatty K (2006a) Oestrogen receptor alpha and beta, androgen receptor and progesterone receptor mRNA and protein localisation within the developing ovary and in small growing follicles of sheep. Reproduction 131:81–92

Juengel J, Reader K, Bibby A, Lun S, Ross I, Haydon L, McNatty K (2006b) The role of bone morphogenetic proteins 2, 4, 6 and 7 during ovarian follicular development in sheep: contrast to rat. Reproduction 131:501–513

Kerr JB, Brogan L, Myers M, Hutt KJ, Mladenovska T, Ricardo S, Hamza K, Scott CL, Strasser A, Findlay JK (2012) The primordial follicle reserve is not renewed after chemical or gamma-irradiation mediated depletion. Reproduction 143:469–476

Kujjo LL, Ronningen R, Ross P, Pereira RJ, Rodriguez R, Beyhan Z, Goissis MD, Baumann T, Kagawa W, Camsari C, Smith GW, Kurumizaka H, Yokoyama S, Cibelli JB, Perez GI (2012) RAD51 plays a crucial role in halting cell death program induced by ionizing radiation in bovine oocytes. Biol Reprod 86(3):76, 1–11

Kurman RJ, Shih I (2010) The origin and pathogenesis of epithelial ovarian cancer: a proposed unifying theory. Am J Surg Pathol 34:433–443

Liew S, Sarraj M, Drummond A, Findlay J (2011) Estrogen-dependent gene expression in the mouse ovary. PLoS One 6:e14672

Matsuda-Minehata F, Inoue N, Goto Y, Manabe N et al (2006) The regulation of ovarian granulosa cell death by pro-and anti-apoptotic molecules. J Reprod Devel 52:695–705

Oktem O, Oktay K (2008) The ovary: anatomy and function throughout human life. Ann N Y Acad Sci 1127:1–9

Pangas S (2007) Growth factors in ovarian development. Semin Reprod Med 25:225–234

Perez GI, Acton BM, Jurisicova A, Perkins GA, White A, Brown J, Trbovich AM, Kim M, Fissore R, Xu J, Ahmady A, D'Estaing SG, Li H, Kagawa W, Kurumizaka H, Yokoyama S, Okada H, Mak TW, Ellisman M, Casper RF, Tilly JL (2007) Genetic variance modifies apoptosis susceptibility in mature oocytes via alterations in DNA repair capacity and mitochondrial ultrastructure. Cell Death Differ 14:524–533

Ratajczak M, Liu R, Marlicz W, Blogowski W, Starzynska T, Wojakowski W, Zuba-Surma E (2011) Identification of very small embryonic/epiblast-like stem cells (VSELs) circulating in peripheral blood during organ/tissue injuries. Methods Cell Biol 103:31–54

Richards J, Pangas S (2010) The ovary: basic biology and clinical implications. J Clin Invest 120:963–972

Rodgers RJ, Irving-Rodgers HF (2010) Morphological classification of bovine ovarian follicles. Reproduction 139:309–318

Roos WP, Kaina B (2012) DNA damage-induced apoptosis: from specific DNA lesions to the DNA damage response and apoptosis. Cancer Lett. Epub ahead of print

Silber SJ (2012) Ovary cryopreservation and transplantation for fertility preservation. Mol Hum Reprod 18:59–67

Spanel-Borowski K (1981) Morphological investigations on follicular atresia in canine ovaries. Cell Tissue Res 214:155–168

Spanel-Borowski K (2010) Weibliche Genitalorgane. In: Zilles K, Tillmann BN (eds) Anatomie. Springer, Heidelberg, pp 546–569

Spanel-Borowski K, Calvo W (1982) Short- and long-term response of the adult dog ovary after 1200 R whole-body X-irradiation and transfusion of mononuclear leukocytes. Int J Radiat Biol Relat Stud Phys Chem Med 41:657–670

Spanel-Borowski K, Thor-Wiedemann S, Pilgrim C (1984) Cell proliferation in the dog (beagle) ovary during proestrus and early estrus. Acta Anat (Basel) 118:153–158

Spanel-Borowski K, Schäfer I, Zimmermann S, Engel W, Adham IM (2001) Increase in final stages of follicular atresia and premature decay of corpora lutea in Insl3-deficient mice. Mol Reprod Dev 58:281–286

Springer LN, Tilly JL, Sipes IG, Hoyer PB (1996) Enhanced expression of bax in small preantral follicles during 4-vinylcyclohexene diepoxide-induced ovotoxicity in the rat. Toxicol Appl Pharmacol 139:402–410

Tavassoli FA, Devilee P (2003) Pathology and genetics of tumours of the breast and female genital organs. International Agency for Research on Cancer, Lyon. Chapter 2, pp 113–202. ISBN 92 832 24 124

Tilly JL (1996) Apoptosis and ovarian function. Rev Reprod 1:162–172

Virant-Klun I, Stimpfel M, Skutella T (2011) Ovarian pluripotent/multipotent stem cells and in vitro oogenesis in mammals. Histol Histopathol 26:1071–1082

Follicular Atresia as a Proliferative and Inflammatory Event

Follicular atresia is considered a process of involution (Hussein 2005). A closer look reveals a considerable number of proliferating cells in regressing preantral and antral follicles (Fig. 3.1) (Spanel-Borowski et al. 1981). In contrast to intact antral follicles, proliferation continues in inner granulosa cells (Fig. 3.1c, d). Like intact follicles, regressing follicles represent the compartment of proliferation, because the cortical stroma lacks any proliferating cells (Fig. 2.2a; Fig. 3.1b, c). Cell renewal of interstitial gland cells thus seems to occur through follicular atresia to a certain degree. Follicle cells, which have escaped cell death, intermingle with the cortical stroma (Fig. 3.2a–c). It signifies tolerance of the cortical stroma against newcomers that are obviously considered as a nonthreatening input. The adjustment probably causes a change in the predominance of meiosis-inhibiting factors in favor of meiosis-activation factors, finally leading to resumption of first meiotic division (Fig. 3.2c, d). Second meiotic division progresses to oocyte cleavages and a morula-like form appears (Fig. 3.2c–f). Oocyte cleavages could explain teratoma formation, a benign ovarian tumor with tissues from one of the three germ layers (Tavassoli and Devilee 2003). Down-regulation of intracellular cyclic adenosine monophosphate levels and degradation of gap junctions between granulosa cells are reported to be involved in the oocyte change from arrest to activation (Jamnongjit and Hammes 2005). It obviously occurs independently of FSH and LH, which underlines the major importance of intrafollicular factors such as amphiregulin, sterols, and steroids for oocyte activation. Large antral follicles under regression provoke heavy leukocyte recruitment, capillary sprouting, and connective tissue hyalinization similar to intact antral follicles (Spanel-Borowski et al. 1997). The signs of inflammation indicate tissue defense that is likely under INIM control (Fig. 3.3). The same holds true for the increase of eosinophils in the theca of regressing antral follicles compared to intact ones (Fig. 3.4). Inflammatory changes coincide with up-regulation of the hypoxia-inducing factor 1α and KIT, the tyrosine kinase growth receptor CD117. The former is responsible for VEGF production and the latter for cell differentiation and migration (Mukhopadhyay and Datta 2004; Rönnstrand 2004).

It appears that regression of follicles compares with similar mechanisms found in preovulatory follicles (Chap. 5). Meiosis is resumed, the basement membrane is disintegrated in conjunction with intraovarian oocyte release (IOR) (Chap. 7), and a physiological inflammation occurs. Seeking a similar molecular network for follicular atresia and for the ovulatory event, a promising candidate is oxidative stress as a danger signal. It generates oxLDL, which is recognized by PRRs on granulosa cell subtypes (Serke et al. 2009) leading to the pro- and anti-inflammatory INIM pathway either in intact or in aging follicles. It is suggested that follicle aging depends on damage by oxidative stress, and that the final steps of the ovulatory event represent a danger-signaling event of INIM (Tatone et al. 2008; Spanel-Borowski 2011a, b). Different degrees of oxidative stress might decide on the resumption of meiosis and/or the strong inflammatory response in follicular atresia.

Granulosa cells are heterogeneous in morphology and in function, also shown for cell cultures (Spanel-Borowski and Ricken 1997). When comparing different species, granulosa cell subtypes appear in preantral and antral follicles under atresia (Fig. 3.5). Important to note are darkly stained granulosa cells in canine antral follicles that form a network through long, branching cell processes (Fig. 3.5a, b). This preferred position in the outer granulosa cell layer is also found for albumin-positive granulosa cells in rat antral follicles (Fig. 3.5c, d). In preantral follicles of golden hamsters, oocyte-associated granulosa cells show an albumin immunoresponse. The appearance of dark granulosa cells and of albumin-positive cells could indicate early atresia. This possibility is deduced from the potential analogy with a CK-positive subtype, which is found in advanced atresia: The CK-positive subtype is localized in the oocyte-associated granulosa cell layer in bovine preantral follicles (Fig. 3.5e, f) and, in antral follicles, CK-positive cells are seen in the

K. Spanel-Borowski, *Atlas of the Mammalian Ovary*,
DOI 10.1007/978-3-642-30535-1_3, © Springer-Verlag Berlin Heidelberg 2012

cumulus and in residues of the outer granulosa (Fig. 3.5g, h). It is noteworthy that another remarkable similarity is becoming evident for structures of follicular atresia and of the ovulatory period. Regressing follicles and preovulatory follicles both display granulosa cell subtypes with CK and with albumin (Chaps. 3 and 10). As described, CK-positive granulosa cell cultures from preovulatory follicles up-regulate TLR4 under oxLDL treatment (Serke et al. 2009, 2010). For activation of the TLR family, heterodimerization with membrane receptors such as CD14 and CD 36 is needed (Miller et al. 2003; Stewart et al. 2010). Provided that coregulation of TLR4 with the albumin receptor is realistic,

albumin could be internalized in a receptor complex by CK-positive granulosa cells. We speculate, owing to the comparable localization, that one subtype resembles the other (Chap. 10). Co-staining techniques can give evidence of whether the subtypes with albumin or with CK are alike. In the final outcome, the subtypes might turn out to be a dendritic-like cell type with a danger-sensing function in intact and regressing follicles.

3.1　Cell Proliferation in the Regressing Follicle Compartment

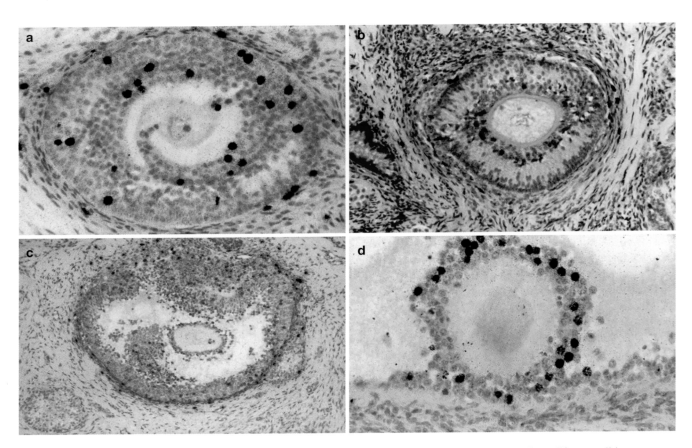

Fig. 3.1 Follicular atresia is a proliferative event as noted in autoradiographs after ³H-thymidine pulse labeling of mature beagles. (**a–d**) Regressing preantral (**a**, **b**) and antral (**c**, **d**) follicles continue with cell proliferation (black grains on nuclei) according to the ³H-thymidine incorporation. The inner granulosa cells proliferate well in contrast to intact follicles (see Fig. 2.2c, d) Cell proliferation in the cortical stroma is poor (**a–c**) (Adapted from Spanel-Borowski et al. 1981) x a,d:230; b:150; c:100

3.2   Transformation into the Interstitial Cortical Tissue and Resumption of Meiosis

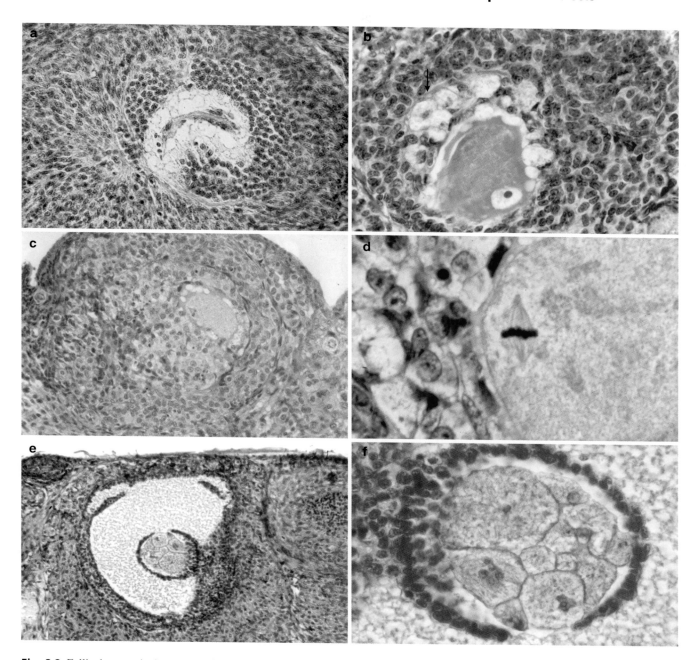

Fig. 3.2 Follicular atresia is connected with cell renewal of the interstitial cortex and with resumption of meiosis as seen in ovaries of a fertile beagle (**a**) and of cyclic rats (**b–f**). (**a**) Thecal cells of a pre-antral follicle blend into the interstitial cortical tissue (*left side*), and the basement membrane between the granulosa and the theca disappears (*right side*). (**b**) A dying oocyte (eosinophilic ooplasm and deformation) is noted in a former preantral follicle. A basement membrane residue separates granulosa cells with fatty degeneration (*arrow*) from interstitial gland cells. They seem to have originated from thecal cells. (**c, d**) The oocyte with a mitotic spindle indicates resumption of meiosis in advanced follicular atresia. Former follicle cells have become interstitial gland cells. (**e, f**) The cleavage of the oocyte in a regressing antral follicle resembles a morula-like structure. HOPA (**a**), H&E (**b-f**), x a,c:200; b,f:450; d:680; e:120

3.3   Inflammatory Responses in Antral Follicle Atresia

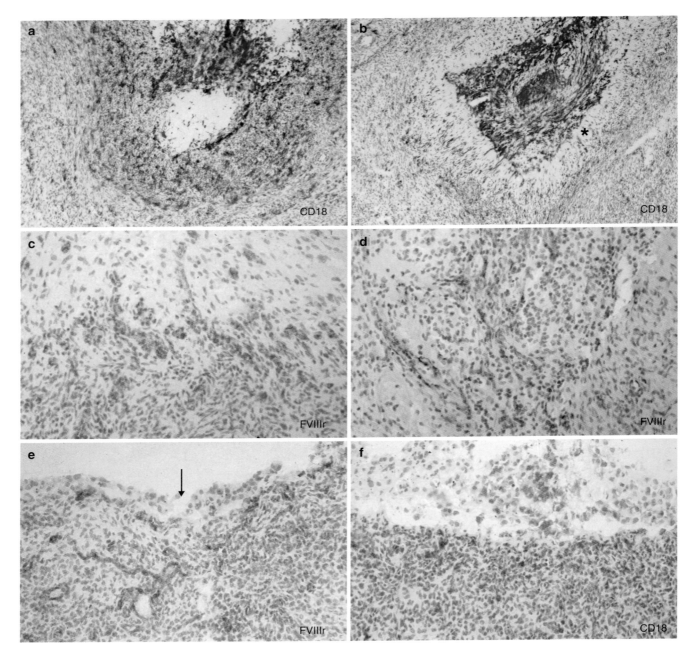

Fig. 3.3 Atresia is an inflammatory process in large antral follicles of the bovine ovary. Immunostaining for CD11/18-positive leukocytes in **a**, **b**, and **f** and for FVIIIr-positive endothelial cells in **c**, **d**, and **e**. (**a**, **b**) Large antral follicles with hypertrophy of the theca are heavily invaded by leukocytes. At the onset of tissue reorganization, leukocytes accumulate in the former follicle wall in **a**. After its hyalinization (*asterisk*), leukocytes are crowded in the former antrum in **b**. (**c**, **d**) Capillaries sprout into the hyalinized follicle wall in **c** and invade the antrum in **d**. (**e**, **f**) Antral follicles without hypertrophy of the theca depict capillary sprouting and leukocytes. The granulosa is reduced in layers (*arrow* in **e**) and has detached during the embedding procedure in **f**. x a,b.90; c,d,e,f:170 (Adapted from Spanel-Borowski et al. 1997)

3.4   Suggested Immunoresponses of the Theca

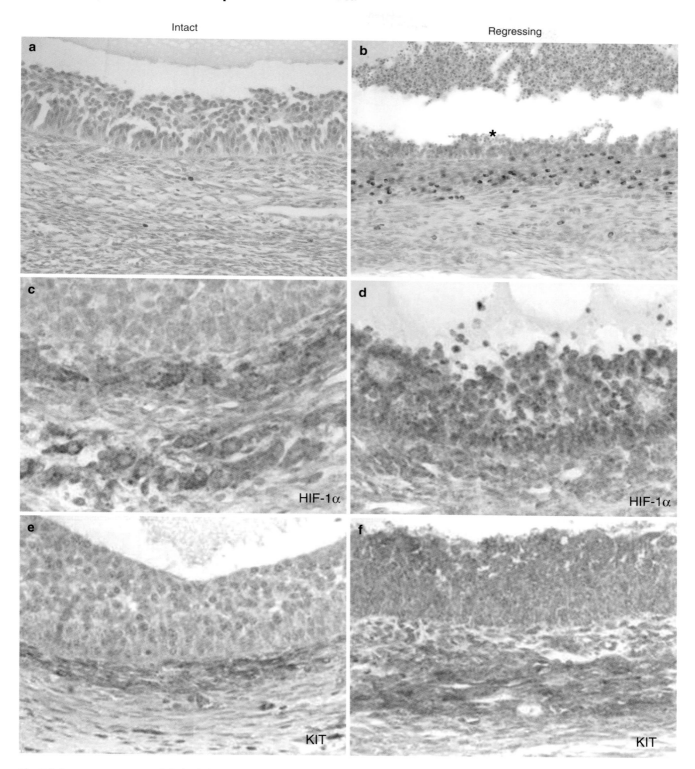

Fig. 3.4 Immunoresponses are inferior in intact antral follicles (*left column*) compared to regressing follicles (*right column*). Cyclic ovaries are from sheep in **a** and **b** (with Sirius-red staining), from rabbits in **c** and **d** (with immunostaining for hypoxia-inducing factor 1α, HIF-1α), and from cows in **e** and **f** (with immunostaining for KIT). (**a**, **b**) Many eosinophils are recruited into the wall of a regressing antral follicle. The granulosa with apoptotic cells is disrupted (*asterisk*). (**c**, **d**) The strong expression of HIF-1α in the theca of an intact antral follicle (**c**) indicates angiogenic activity. Interstitial gland cells in the lower part are also positive. The strong expression of HIF-1α in the granulosa and the theca of a regressing antral follicle in **d** speaks for angiogenic activity in both layers. (**e**, **f**) KIT positivity is up-regulated in the granulosa and the theca of regressing antral follicles. x a,b:130; c,d,e,f:260 (Images are from the MD thesis of Stefan Karger, Corinna Meyer, and Winnie Kunzemann, all from Leipzig)

3.5   Granulosa Cell Subtypes

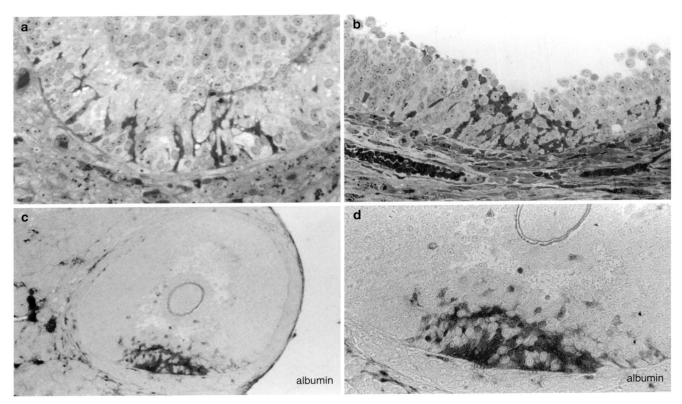

Fig. 3.5 Granulosa cell subtypes appear during follicular atresia in ovaries of dogs (**a–d** for Richardson-stained semithin sections), of adult golden hamsters (**c–f** for immunohistology of albumin localization in brown without H&E counterstaining), and of cows (**f–h** for immunostaining of cytokeratin in red, CK). Albumin-positive cells could compare with dark cells and with CK-positive granulosa cells. (**a**) The dark cells of the large preantral follicle (Fig. 2.4f) form long interacting processes in the basal cell layers. (**b**) Dark-blue cells form a network in the mural granulosa cell layer of an antral follicle (**c, d**). Albumin-positive cells (*brown*) of columnar shape in a large preantral follicle are reminiscent of the dark cells in **a**. The deformation of the oocyte signifies the onset of atresia. (**e**) Many granulosa cells adjacent to the zona pellucida are albumin-positive, which might indicate the suggested rupture of the zona pellucida as a sign of early atresia. (**f**) The strong upregulation of CK in the inner granulosa cell layer of a preantral follicle (CK-positive cells in *red*) is associated with deformations of the zona pellucida and the oocyte according to pathway A of follicular atresia. (**g**) CK-positive cells of fibroblast and epithelioid type are noted in the cumulus oophorus and mural granulosa cells of a cystic follicle. (**h**) Mural granulosa cells of a cystic follicle form two layers of epithelioid cells with strong CK positivity. A single CK-positive cell is apparent in the stroma (*arrow*). x a:330; b,d:260; c:120; e,h:270; f:200; g:80

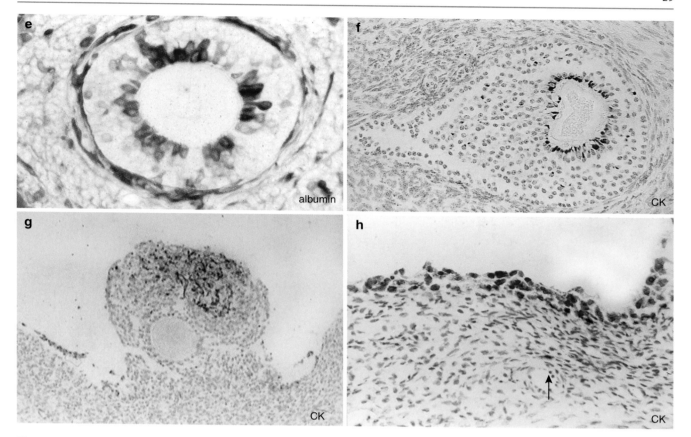

albumin

CK

CK

CK

Fig. 3.5 (continued)

References

Hussein MR (2005) Apoptosis in the ovary: molecular mechanisms. Hum Reprod Update 11:162–177

Jamnongjit M, Hammes S (2005) Oocyte maturation: the coming of age of a germ cell. Semin Reprod Med 23:234–241

Miller YI, Viriyakosol S, Binder CJ, Feramisco JR, Kirkland TN, Witztum JL (2003) Minimally modified LDL binds to CD14, induces macrophage spreading via TLR4/MD-2, and inhibits phagocytosis of apoptotic cells. J Biol Chem 278:1561–1568

Mukhopadhyay D, Datta K (2004) Multiple regulatory pathways of vascular permeability factor/vascular endothelial growth factor (VPF/VEGF) expression in tumors. Semin Cancer Biol 14:123–130

Rönnstrand L (2004) Signal transduction via the stem cell factor receptor/c-Kit. Cell Mol Life Sci 61:2535–2548

Serke H, Bausenwein J, Hirrlinger J, Nowicki M, Vilser C, Jogschies P, Hmeidan FA, Blumenauer V, Spanel-Borowski K (2010) Granulosa cell subtypes vary in response to oxidized low-density lipoprotein as regards specific lipoprotein receptors and antioxidant enzyme activity. J Clin Endocrinol Metab 95:3480–3490

Serke H, Vilser C, Nowicki M, Hmeidan FA, Blumenauer V, Hummitzsch K, Lösche MT, Spanel-Borowski K (2009) Granulosa cell subtypes respond by autophagy or cell death to oxLDL-dependent activation of the oxidized lipoprotein receptor 1 and toll-like 4 receptor. Autophagy 5:991–1003

Spanel-Borowski K (2011a) Footmarks of innate immunity in the ovary and cytokeratin-positive cells as potential dendritic cells, vol 209, Advances in anatomy, embryology, and cell biology. Springer, Heidelberg. ISBN 978-3-642-16076-9

Spanel-Borowski K (2011b) Ovulation as danger signaling event of innate immunity. Mol Cell Endocrinol 333:1–7

Spanel-Borowski K, Ricken AM (1997) Varying morphology of bovine granulosa cell cultures. In: Motta M (ed) Microscopy of reproduction and development: a dynamic approach. Antonio Delfino, Rome, pp 91–100

Spanel-Borowski K, Trepel F, Schick P, Pilgrim C (1981) Aspects of cellular proliferation during follicular atresia in the dog ovary. Cell Tissue Res 219:173–183

Spanel-Borowski K, Rahner P, Ricken AM (1997) Immunolocalization of CD18-positive cells in the bovine ovary. J Reprod Fertil 111:197–205

Stewart CR, Stuart LM, Wilkinson K, van Gils J, Deng J, Halle A, Rayner KJ, Boyer L, Zhong R, Frazier WA, Lacy-Hulbert A, Khoury J, Golenbock DT, Moore KJ (2010) CD36 ligands promote sterile inflammation through assembly of a toll-like receptor 4 and 6 heterodimer. Nat Immunol 11:155–161

Tatone C, Amicarelli F, Carbone MC, Monteleone P, Caserta D, Marci R, Artini PG, Piomboni P, Focarelli R (2008) Cellular and molecular aspects of ovarian follicle ageing. Hum Reprod Update 14:131–142

Tavassoli FA, Devilee P (2003) Pathology and genetics of tumours of the breast and female genital organs. International Agency for Research on Cancer, Lyon. Chapter 2, pp 113–202, ISBN 92 832 241 24

The Cortex and Cellular Stromatolysis

The architecture of the ovarian cortex is widely neglected as regards specific functions. The surface epithelium is an exception, because of its relationship to the origin of epithelial ovarian cancer (Auersperg et al. 2001; Kurman and Shih 2010). The one-epithelial-layer surface cells are anchored in the tunica albuginea, a thickened basement membrane about 150 μm in width in the cow (Vigne et al. 1994). The tunica albuginea zone additionally comprises surface stromal cells of perpendicular and parallel layers (Fig. 4.1a–d). It is noteworthy that the commonly used homeobox genes such as glyceraldehyde 3-phosphate dehydrogenase and β-actin are not expressed when tissue homogenates from the tunica albuginea zone are used for internal control in molecular analysis (unpublished findings). Adjacent to the tunica albuginea zone, the interstitial cortical stroma separates follicles and CLs. The outer/peripheral zone of the bovine cortical stroma hosts primordial, primary, and a few small preantral follicles in a vessel-poor zone about 400 μm thick in the cow (Fig. 4.1b) (Herrman and Spanel-Borowski 1998). The remarkable device keeps the immature follicle stages in a nutrition-poor environment similar to resting progenitor cells in a bone-marrow niche. Accidental contact with a capillary loop could overcome inhibitory transcription factors such as Foxo3 and could activate nerve growth factors and the neurotrophin kinase receptors to launch the primordial follicle into its career (Kerr et al. 2009; Oktem and Urman 2010). In the outer bovine cortical stroma, the density of mast cells is far greater than that of MHC II-positive leukocytes (Fig. 4.1c, d). In the inner zone between 200 and 600 μm in width, the total number of leukocytes is revealed by CD11/18 antigen immunolocalization. The density exceeds the sum of mast cells and MHC II-positive leukocytes in the outer cortical stroma. This finding coincides with a well-developed microvascular bed derived from medullary vessels (Fig. 4.1b). It could support development of antral follicles during their ingrowth from the outer into the inner zone. Additionally, the inner zone considered as a transition zone toward the medulla represents a favorable site of recruitment of leukocytes into the cortical stroma. Because leukocyte numbers are highest in the transition zone (unpublished data), endothelial cell adhesion molecules are most likely different in expression densities when comparing the outer and inner zone. When considering a potential INIM control through mast cells, the tunica albuginea zone is widely excluded because of the absence of mast cells. They are indispensable soldiers that can amplify or suppress innate or acquired immunoresponses (Galli et al. 2011; Shelburne and Abraham 2011). In the bovine ovary, numerous mast cells appear in the outer and inner cortical stroma with the exception of follicles and of CLs (Figs. 4.1b, 5.11f, 6.6d). The mission of mast cells could be to avoid an inflammatory battlefield, because the interstitial cortical stroma is continuously exposed to the danger of tissue damage by follicular growth and atresia.

The interstitial cortical stroma contains the interstitial gland cells as primarily androgen-producing cells (Guraya 1978). Because androgen is aromatized to estrogen by aromatase-producing granulosa cells, the interstitial gland cells indirectly promote follicle growth by delivery of the estrogen substrate. The function thus compares with interstitial/Leydig cells of the testis, which are important for spermatogenesis. Interstitial gland cells of the ovary mature under LH stimulation. They can look like epithelioid endocrine cells and increase steroidogenesis, which intensifies follicle growth. In photosensitive rodents such as the white-footed mouse kept under long photoperiods, interstitial gland cells are mature (Spanel-Borowski et al. 1983). With the decrease of hypothalamic LH secretion under short photoperiods, mature interstitial gland cells become quiescent with a conspicuous decrease of cell size and function (Fig. 4.2a, b). In hypophysectomized rats, the inactive interstitial gland cells are called deficiency cells, which can be stimulated to full size by LH application (Carithers and Green 1972). The occurrence of one or the other cell type depends on the endocrine status of the ovarian cycle and on the species (Guraya 1978). In the adult beagle, inactive interstitial gland cells align in between ribbons of mature interstitial gland cells or both cell types are intermingled (Fig. 4.2c, d). After whole-body X-irradiation of adult beagles, the inactive interstitial cell type is seen with radiation effect I and the mature type is seen with radiation effect II (compare Fig. 4.2e and f with

K. Spanel-Borowski, *Atlas of the Mammalian Ovary*,
DOI 10.1007/978-3-642-30535-1_4, © Springer-Verlag Berlin Heidelberg 2012

Figs. 2.8a and 2.9a) (Spanel-Borowski and Calvo 1985). The interstitial gland cell pool comprises immature, mature, and quiescent cells (Motta et al. 1971; Mossman and Duke 1973; Guraya 1978). Interstitial gland cells are terminally differentiated cells, and part of the renewal system seems to come from follicle cells surviving atresia by transformation into cortical cells (Fig. 3.2a–c). Single KIT-positive progenitor cells of rare occurrence in the cortical stroma (Spanel-Borowski et al. 2007; Honda et al. 2007) are certainly another option for the origin of immature interstitial gland cells. By developing gonadotropic receptors, they could become a new input for the interstitial gland cell pool. In the absence or presence of LH, interstitial gland cells remain either inactive/quiescent or they become mature, respectively.

The cyclic change between the quiescent cell type and the mature type is conceived as an appealing design from the viewpoint of "form follows function." During the adaptation process from a mature to a quiescent form, the mature cell type sheds the peripheral cytoplasm together with organelles into the intercellular stroma (Fig. 4.3). A new cellular membrane is reconstituted with cytoplasmic residues close to the nucleus. It shrinks and increases the heterochromatin as a sign of inactivation (Fig. 4.3f, g). The cytoplasmic shedding of mature interstitial gland cells is termed "cellular stromatolysis" because the uniform decrease in the interstitial gland cell size significantly correlates with the decrease in width of the interstitial cortical stroma and also of the ovarian weight (Spanel-Borowski et al. 1983; Spanel-Borowski and Calvo 1985). Cytoplasmic shedding has long been known for interstitial gland cells in the mammalian ovary (Motta et al. 1971; Guraya 1978). A similar process in megakaryocytes gives rise to thrombocytes or in epithelial cells to apocrine secretion (Bloom and Fawcett 1975). Extracellular microvesicles

occur through ectocytosis/microvesicle shedding of plasma membrane blebs associated with micrometer-sized membrane defects (McNeil 1993; Herrman et al. 1996; Cocucci et al. 2009; Mathivanan et al. 2010). Growth factors such as the basic fibroblast factor without signal sequences are released into the extracellular space by microvesicle shedding. These examples show evidence of complete cell repair after transient plasma membrane defects. Cytoplasmic shedding in mature interstitial gland cells might be considered as amplified ectocytosis to rapidly lower high function to a basic level by avoiding cell death.

The phenomenon of cytoplasmic shedding should be addressed as a novel way of cell inactivation/degeneration. When a stimulating endocrine and paracrine microenvironment is re-established, inactivated/quiescent interstitial gland cells are able to completely recover. The speculated cycle of cytoplasmic shedding and cell recovery keeps interstitial gland cells alive as long as their life span allows. As is generally accepted today, auto- and heterophagocytosis with the appearance of autophagosomes, myelin bodies, apoptotic bodies, and macrophages characterize the pathology of cell degeneration. These morphological markers might accompany cytoplasmic shedding. Altogether, the life cycle of interstitial gland cells (Fig. 4.4) awaits further study, including investigation of whether the release of cytoplasmic components is responsible for the presence of CD18-positive leukocytes in the cortical stroma (Fig. 4.1c, e). More than 70 % of CD18-positive cells are mast cells in the bovine ovary, and the number does not change significantly during the estrous cycle (Reibiger and Spanel-Borowski 2000). Mast cells contain a plethora of powerful factors that can amplify or inhibit INIM processes (Galli et al. 2011; Shelburne and Abraham 2011).

Fig. 4.1 Zones in the cortex and medulla are depicted for the bovine ovary after immunostaining for FVIIIr-positive endothelial cells (**a, b**), for mast cells (toluidine blue staining in **c**), and for MHC II- and CD18-positive for leukocytes (**d, e**). (**a**) The micrograph of mounted images displays the surface epithelium and the ovarian surface stromal cells in the very periphery. Zone I of the interstitial cortical stroma consists of interstitial gland cells and of mainly quiescent follicles. Zone II represents the transition zone between the cortex and the medulla. Microvessels become prominent in zone II, in which large follicles extend (Fig. 2.1a). The medulla shows larger microvessels, shunt-like vessels, and lymph vessels. (**b**) The surface epithelium zone and cortical zone I with primordial and primary follicles are sparsely vascularized. (**c, d**) Mast cells and MHC II-positive leukocytes are negligible in the surface epithelium zone. (**e**) Many CD18-positive leukocytes populate cortical zone II with a hyalinized antral follicle (*asterisk*). The density of CD18-positive leukocytes comprising mast cells, segmented and mononuclear leukocytes, is higher than that of mast cells and MHC II-positive leukocytes. x a,b,c,d.80 (**a, d** – From the MD thesis of Winnie Kunzemann) (Adapted from Herrman and Spanel-Borowski 1998; Reibiger and Spanel-Borowski 2000)

4.1   Architecture of the Cortex

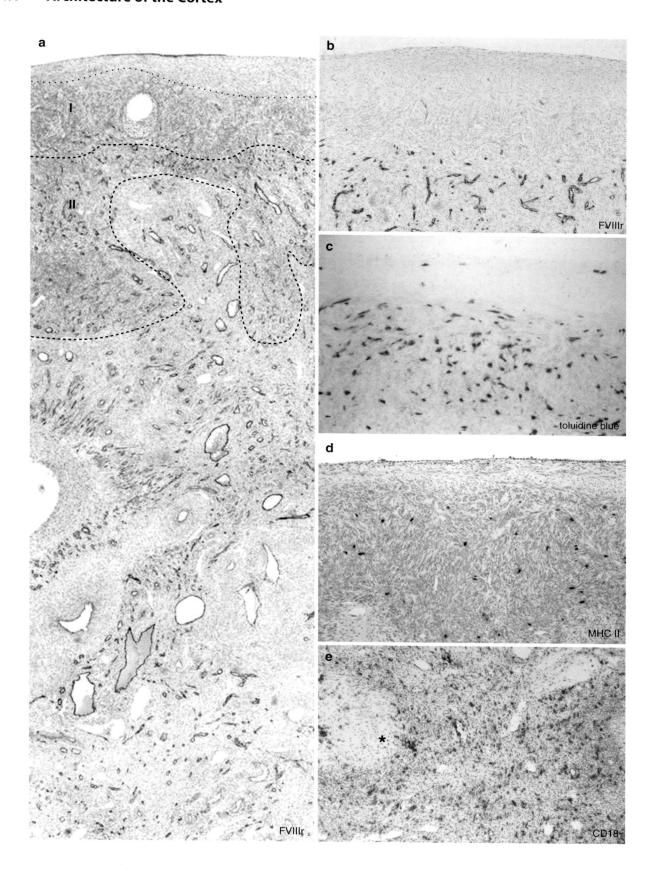

4.2   Interstitial Gland Cell Types

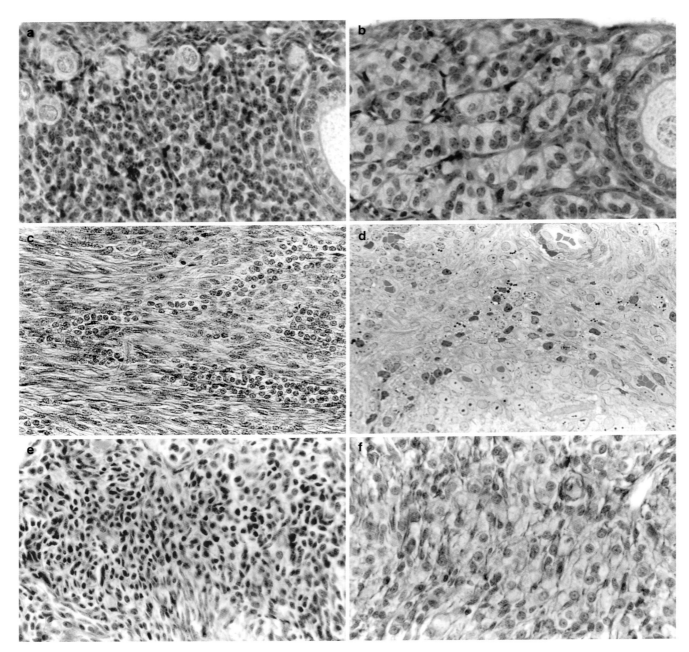

Fig. 4.2 The interstitial cortical tissue consists of resting/quiescent/ inactive or mature interstitial gland cells in ovaries of white-footed mice (*Peromyscus leucopus*) in (**a** and **b**) (H&E staining) and in beagles in (**c**, **e**, and **f**) (HOPA staining) and (**d**) (Richardson-stained semithin section). (**a**, **b**) The quiescent cell type occurs under a short photoperiod (**a**) in comparison to the mature epithelioid type under a long photoperiod (**b**). (**c**) Mature interstitial gland cells show a radial arrangement between stroma-like elements. The mature cells possess a vesicular nucleus and are rich in cytoplasm. (**d**) The inactive cell type with hyperchromatic irregular nuclei and sparse cytoplasm intermingles with the mature cell type. (**e**, **f**) 1,200-R (12 Gy) whole-body X-irradiation of adult beagles together with transfusion of mononuclear cells causes either quiescence or hypertrophy/activation of interstitial gland cells with radiation effect I or II (Figs. 2.8 and 2.9). x a,b,c,e,f:240;d.300 (Adapted from Spanel-Borowski et al. 1983; Spanel-Borowski and Calvo 1985)

4.3   Cytoplasmic Shedding

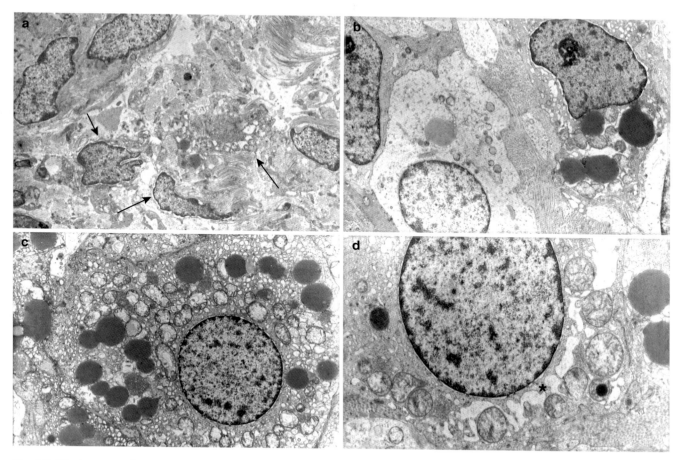

Fig. 4.3 The immature, the mature, and the inactive interstitial gland cell types are seen in the process of cytoplasmic shedding on ultrathin sections of the beagle ovary. (**a**) The overview depicts resting gland cells of different stages (*arrows*) and shows abundant cellular debris in the extracellular space. (**b**) The immature cell with a round nucleus and a small amount of condensed chromatin displays inconspicuous organelles similar to an immature granulosa cell (*left side*). The right-sided cell becomes quiescent by cytoplasmic shedding. (**c**) The mature gland cell represents a steroid-secreting cell with tubular mitochondria, lipid droplets, and smooth endoplasmic reticulum. (**d**) The mature type at the onset of the resting event develops a demarcation channel (*asterisk*). (**e**) The early stage of cytoplasmic shedding is characterized by a lace-like cytoplasm. (**f, g**) The advanced stage displays shedding of peripheral cytoplasm and restoration of a new cell membrane (*arrow*). The inactive cell shows a prominent nucleolus in a shrunken nucleus with condensed chromatin. (**h**) Vesicles and a cell remnant (*arrow*) appear among collagen fibers of the cortical stroma x a:2'800; b:5'000; c:5'500; d:8'200; e:7'200; f:'7200; g:9'300; h:19'500 (Adapted from Spanel-Borowski and Calvo 1985)

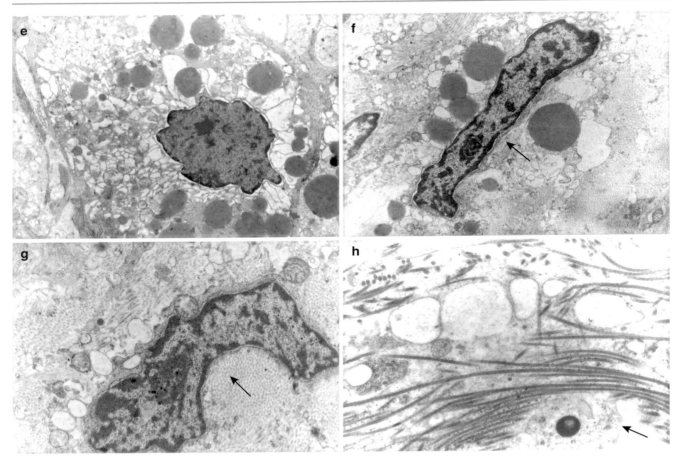

Fig. 4.3 (continued)

4.4 Scheme for Life Cycle of Interstitial Gland Cells

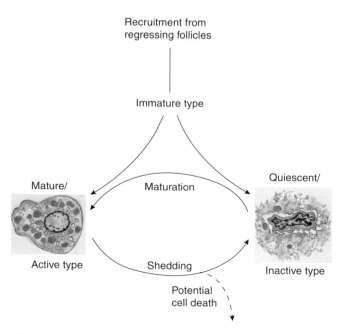

Fig. 4.4 The life cycle of interstitial gland cells starts with immature interstitial gland cells, which are recruited from regressing follicles. The immature cell type becomes the mature type or the quiescent type depending on the presence or absence of receptor-dependent stimulation. Stimulation of the quiescent cell type transforms it into the mature type. Conversely, without stimulation, the mature type undergoes cytoplasmic shedding leading to the quiescent cell type. Uncontrolled shedding can cause cell death (Drawn by R. Spanel)

References

Auersperg N, Wong A, Choi KC, Kang SK, Leung PC (2001) Ovarian surface epithelium: biology, endocrinology, and pathology. Endocr Rev 22:255–288

Bloom W, Fawcett D (1975) A textbook of histology, 10th edn. Saunders, Philadelphia, London

Carithers JR, Green JA (1972) Ultrastructure of rat ovarian interstitial cells. I. Normal structure and regressive changes following hypophysectomy. J Ultrastruct Res 39:239–250

Cocucci E, Racchetti G, Meldolesi J (2009) Shedding microvesicles: artefacts no more. Trends Cell Biol 19:43–51

Galli S, Borregaard N, Wynn T (2011) Phenotypic and functional plasticity of cells of innate immunity: macrophages, mast cells and neutrophils. Nat Immunol 12:1035–1044

Guraya SS (1978) Recent advances in the morphology, histochemistry, biochemistry, and physiology of interstitial gland cells of mammalian ovary. Int Rev Cytol 55:171–245

Herrman G, Spanel-Borowski K (1998) A sparsely vascularised zone in the cortex of the bovine ovary. Anat Histol Embryol 27:143–146

Herrman G, Missfelder H, Spanel-Borowski K (1996) Lectin binding patterns in two cultured endothelial cell types derived from bovine corpus luteum. Histochem Cell Biol 105:129–13

Honda A, Hirose M, Hara K, Matoba S, Inoue K, Miki H, Hiura H, Kanatsu-Shinohara M, Kanai Y, Kono T, Shinohara T, Ogura A (2007) Isolation, characterization, and in vitro and in vivo differentiation of putative thecal stem cells. Proc Natl Acad Sci USA 104:12389–12394

Kerr B, Garcia-Rudaz C, Dorfman M, Paredes A, Ojeda S (2009) NTRK1 and NTRK2 receptors facilitate follicle assembly and early follicular development in the mouse ovary. Reproduction 138:131–140

Kurman RJ, Shih I (2010) The origin and pathogenesis of epithelial ovarian cancer: a proposed unifying theory. Am J Surg Pathol 34:433–443

Mathivanan S, Ji H, Simpson R (2010) Exosomes: extracellular organelles important in intercellular communication. J Proteomics 73:1907–1920

McNeil PL (1993) Cellular and molecular adaptations to injurious mechanical stress. Trends Cell Biol 3:302–307

Mossman H, Duke K (1973) Comparative morphology of the mammalian ovary. Univ. Madison Wisconsin Press, Madison

Motta P, Nesci E, Fumagalli L (1971) The fine structure and cyclic morphologic changes of the interstitial cells in the mammalian ovary. Arch Anat Histol Embryol 54:43–58

Oktem O, Urman B (2010) Understanding follicle growth in vivo. Hum Reprod 25:2944–2954

Reibiger I, Spanel-Borowski K (2000) Difference in localization of eosinophils and mast cells in the bovine ovary. J Reprod Fertil 118:243–249

Shelburne C, Abraham S (2011) The mast cell in innate and adaptive immunity. Adv Exp Med Biol 716:162–185

Spanel-Borowski K, Calvo W (1985) Degeneration of interstitial gland cells of normal and irradiated dog ovary. Arch Anat Microsc Morphol Exp 74:167–177

Spanel-Borowski K, Petterborg LJ, Reiter RJ (1983) Morphological and morphometric changes in the ovaries of white-footed mice (peromyscus leucopus) following exposure to long or short photoperiod. Anat Rec 205:13–19

Spanel-Borowski K, Sass K, Löffler S, Brylla E, Sakurai M, Ricken AM (2007) KIT receptor-positive cells in the bovine corpus luteum are primarily theca-derived small luteal cells. Reproduction 134:625–634

Vigne JL, Halburnt LL, Skinner MK (1994) Characterization of bovine ovarian surface epithelium and stromal cells: identification of secreted proteins. Biol Reprod 51:1213–1221

Rupture of the follicle wall is the culminating point in the ovulatory period. The diverse theories on the rupture mechanism—such as an increase in intrafollicular pressure, activation of metalloproteinases and plasminogen activator (Fig. 5.1), inflammatory-like response and contraction of myofilament-positive thecal cells—are united by the concept of INIM as the final driving force (Spanel-Borowski 2011a, b). As is assumed, the ovulatory period is initiated by the LH peak and continued by oxidative stress in the preovulatory follicle. Oxidative stress is perceived as danger by INIM, which responds with a proinflammatory signaling cascade regulated by CK-positive granulosa cells with TLR4 expression (Serke et al. 2009, 2010). An inside-out process terminates connective tissue degradation and oocyte release. Oxidative stress in the preovulatory follicle depends on an increased production of reactive oxygen species (ROS) as byproducts of full-speed steroidogenesis. ROS are released from the mitochondrial respiratory chain through leaky membranes (Hanukoglu 2006) and also produced by an oxLDL-dependent activation of the lectin-like oxidized low-density lipoprotein receptor (LOX-1), which is expressed by granulosa cells (Serke et al. 2009; Vilser et al. 2010). Via a vicious internal feedback loop, oxLDL binding increases LOX-1 expression and ROS production (Mehta 2006; Chen 2007). In minor concentrations, ROS might be beneficial and mediate resumption of meiosis (Fig. 5.2), because reactive oxygen can function as a second messenger in the signal transduction pathway of INIM signaling (Kohchi et al. 2009). In higher concentrations, however, ROS seem to be detrimental and account for 20–50 % of dead granulosa cells in fresh follicle harvests of women under IVF therapy (Vilser et al. 2010). Norepinephrine uptake and metabolism were recently found to be another inducer of ROS generation in granulosa cells of preovulatory follicles (Saller et al. 2012).

ROS oxidize locally available lipoproteins to oxLDL, which binds to TLR4 of CK-positive granulosa cells (Serke et al. 2009, 2010). When the subsequent TLR4 signaling cascade chooses the myeloid differentiation adaptor factor 88 (Myd88) gateway, proinflammatory changes such as

angiogenesis take place (O'Neill and Bowie 2007; Takeuchi and Akira 2010). Early steps of capillary sprouting are nicely displayed in microvessel corrosion casts of superovulated hamster ovaries (Löseke and Spanel-Borowski 1996). Six hours after application of LH, 18 out of 20 preovulatory follicles display sprouts in concentric orientation below the basement membrane, whereas sprouts are radially aligned 38 h later (Fig. 5.3). Hence, granulosa cells seem to communicate with thecal microvessels to induce sprouts before follicle rupture. Sprouts are kept waiting below the basement membrane until the moment of rupture. This triggers ingrowth of sprouts from the vascular theca into the avascular granulosa cell layer. The follicle apex contains the oocyte bed either as a sinusoid, as a defect in the corrosion casts, or as an oocyte replica (Fig. 5.4). A time sequence is suggested for the three structures: The sinusoid develops before the rupture of a preovulatory follicle, most likely to supply the oocyte area with sufficient blood. The punched-out defect of oocyte size gives an idea of the form and extension of the rupture site (Fig. 5.4a, b). The replica seems to appear when the oocyte-associated sinusoid becomes permeable to generate the circumscribed preovulatory edema and hemorrhage around the mature oocyte (Byskov 1969; Bjersing and Cajander 1974). The replica is to be understood as an artifact obtained by extravasation of resin at the moment of vessel casting. The oocyte is entrapped at the rupture site in the resin leakage and there it imprints its body as a cap-like form (Fig. 5.4c, d). The cap-like form in the CL (Fig. 5.4e, f) thus alludes to retention of the oocyte, known as unruptured follicle syndrome (Qublan et al. 2006).

Ovaries can be severely harmed by repeated gonadotropic stimulations. They are performed in golden hamsters as soon as CLs have disappeared and the next ovarian cycle starts (Löseke and Spanel-Borowski 1996). Hypo-ovulation occurs, as deduced from ovaries with more persistent preovulatory follicles than CLs (Fig. 7.12b). The cap-like oocyte replica is not apparent in microvessel corrosion casts after repeated superovulations. Severe failures of thecal microvessels are reflected by the thin and dilated thecal capillaries with high

K. Spanel-Borowski, *Atlas of the Mammalian Ovary*,
DOI 10.1007/978-3-642-30535-1_5, © Springer-Verlag Berlin Heidelberg 2012

vessel permeability and by defects of the two-layered thecal vessels (Fig. 5.5). Increased secretion of VEGF and angiopoietin-3 could lead to vessel permeability and inadequate maturation of the preovulatory microvascular bed. The angiogenic pathway is part of the TLR4 signaling cascade over the Myd88 adaptor protein (O'Neill and Bowie 2007; Takeuchi and Akira 2010). Hence, hyperactivation of INIM function seems to be involved in the microvessel changes after repeated stimulations of the ovary and explains the ovarian hyperstimulation syndrome in women undergoing IVF fertilization (Kahnberg et al. 2009). Practical experience has shown that there is a minor effectiveness in IVF outcomes when the 3-month break between two stimulation protocols is disregarded.

Considering the preovulatory follicle as a structure under oxidative stress (Spanel-Borowski 2011b; Devine et al. 2012), the dangerous micromilieu is expected to stimulate protective organelles in the oocytes. A promising candidate is the endoplasmic-reticulum-derived multilamellar body (EMB), an unexpected novel observation made from oocytes of oxLDL-treated mouse follicle cultures as published elsewhere (Spanel-Borowski et al. 2012). Cultures were made by Professor Eichenlaub-Ritter, University of Bielefeld, Germany, according to her own publications (Hu et al. 2001; Sun et al. 2005). Follicles were grown in vitro and reached maturational competence on day 12. Preantral follicles were cultured in the presence of 100 μg/ml oxLDL/nLDL for 20 and 48 h, up to 13 days. Thirteen-day-old cultures were ovulated with 1.5 IU/ml recombinant human chorionic gonadotropin (rhCG) and 5 ng recombinant epidermal growth factor (rEGF), and cumulus–oocyte complexes (COCs) were obtained 18 h later. The EMB compares with a circular body surrounded by a multilamellar envelope (Snapp et al. 2003; Korkhov et al. 2008; Korkhov 2009). The center encloses cytoplasm together with damaged mitochondria and vesicles. The steps of EMB morphogenesis are seen clearly in oxLDL-treated mouse follicles and in COCs (Figs. 5.6, 5.7, and 5.8). Oocytes from preantral follicles contain EMBs with an envelope of rough and smooth ER tubules (type I). Oocytes from COCs organize the envelope from smooth ER vesicles (type II). Controls without lipoprotein treatment lack any EMBs. Yet in the presence of normal low-density lipoprotein (nLDL), the morphogenesis of EMBs appears to be less intense. It is speculated that the EMB is a novel form of autophagy to rapidly remove damaged organelles and proteins. The EMBs could be fragmented into multivesicular bodies and multivesicular complexes. Multivesicular bodies are interpreted as a sign of early atresia in resting follicles and carry the microtubule light chain 3 (LC3) marker, a well-accepted autophagy marker, in mouse zygotes (de Bruin et al. 2002; Klionsky 2007; Tsukamoto et al. 2008).

In preovulatory follicles of canine and bovine ovaries, gyration of the follicle wall occurs. It is caused by a lack of antrum space required by the follicle cells, which increase in number and in size by luteinization. The theca provides infoldings, which infold the granulosa (Fig. 5.9a–c). It is widely overlooked that the reorganization of the preovulatory follicle wall represents physical stress for granulosa cells. Some of them die without appropriate protection (Fig. 5.9b). Before the very moment of follicle rupture, the oocyte detaches from the cumulus oophorus, it floats in the follicular fluid, and undergoes cumulus expansion and mucification of the corona radiata cells (Fig. 5.9c-e). Now the gyrated wall of a preovulatory follicle displays two distinct microvessel layers in the theca (Fig. 5.10a, c, e). The layers heavily recruit CD18-positive leukocytes (Spanel-Borowski et al. 1997). Among them are macrophages that settle in the area of the basement membrane where they likely release metalloproteinases for matrix degradation (Fig. 5.10b, d, f). After the culminating point of ovulation, i.e., of rupture, the two microvessel layers disappear because vessels change in orientation from a concentric alignment into a radial one (Rohm et al. 2002). The basement membrane disintegrates, and capillary buds extend into the granulosa. Immunohistological staining of thecal capillaries nicely displays the orientation change of thecal sprouts as also revealed in microvessel corrosion casts (compare Fig. 5.11a and c with Fig. 5.3c–f). CD18-positive leukocytes uniformly distribute in the follicle wall under heavy luteinization and angiogenesis (Fig. 5.11d–f). It is noteworthy that mast cells are missing in the preovulatory follicle (Fig. 5.11f) (Reibiger and Spanel-Borowski 2000). Mast cells, just like eosinophils and dendritic cells, belong to the specific immunocompetent cells of INIM. The population, which is heterogeneous in localization and function, is known to promote inflammatory responses and also to be able to slow them down (Kalesnikoff and Galli 2008).

The ovulatory period compares with enormous tissue reorganization. The mature follicle is dismissed and residues are taken for reconstruction into a CL, similar to the substantial renovation of a house for new purposes. Not all follicle cells survive the reorganization process, as is deduced from aspirates of preovulatory follicles obtained from women under IVF therapy. The aspirate contains between 20 and 50 % of dead granulosa cells, whereby high percentages are found in women who are obese and of older reproductive age (Fig. 5.12a, b) (Vilser et al. 2010). Both of these parameters are associated with lower catalase levels in the follicular fluid than in women of normal weight and young reproductive age (Bausenwein et al. 2010). Adequate antioxidant concentrations obviously protect against oxidative stress in the ovulatory event (Tatemoto et al. 2004; Tatone et al. 2008; Whitaker and Knight 2008). Misbalances seem to occur in ovarian failures such as the polycystic ovary syndrome and the metabolic syndrome. Different granulosa cell death forms appear to be executed by ROS molecules. In fact, cell death is beyond apoptosis in the follicle cell harvest of obese women. Defective mitochondria with degenerating lamellae and bodies with multilamellar lipid membranes are indicative of ROS-related cell damage (Fig. 5.12c, d). In the early stage of cell death, characteristic autophagosomes with a double unit membrane around organelle debris can be erroneously taken as pentalaminar annular junctions. They represent internalized gap junctions of preovulatory granulosa cells (Spanel-Borowski and Sterzik 1987) (Fig. 5.13a–c). Autophagosomes in still-intact granulosa cells indicate the struggle for survival through reparative autophagy (Klionsky 2007; Klionsky 2009; Chen and Klionsky 2011). When the survival program gets out of control, cell death autophagy ensues. It might be nonapoptotic with lysis of the nucleus and cytoplasm, as well as apoptotic with chromatin condensation and plasma membrane blebbing (Fig. 5.13d–f). Different cell death forms can be verified in granulosa cell culture subtypes (Fig. 5.14). The fibroblast-like CK-negative type cell, which increases LOX-1 expression between 0 and 36 h of oxLDL/nLDL treatment, undergoes reparative autophagy (Serke et al. 2009, 2010). The epithelioid-like CK-positive cells that up-regulate TLR4 together with CD14 show nonapoptotic cell death. In other words, granulosa cells choose various ways to die. Apoptosis is just one way, although it is the most obvious one in antral follicles (Fig. 2.6c, d) (Tilly 1996; Inoue et al. 2011). Evidence of alternative pathways has been given by others (van Wezel et al. 1999).

Preovulatory follicles contain progenitor cells, recognized by the expression of the tyrosine kinase receptor KIT, i.e., the proto-oncogene CD117 (Merkwitz et al. 2010). A few KIT-positive cells apparently migrate from thecal vessels into the preovulatory follicle wall (Spanel-Borowski et al. 2007). It is still a hot debate whether somatic progenitor cells are present (Szotek et al. 2006; Honda et al. 2007; Tilly and Rueda 2008). Progenitor cells could differentiate into endothelial precursor cells and into cells of the monocyte lineage. The outcome can be demonstrated with hematopoietic-like clones, which develop in long-term bovine granulosa cell cultures (Spanel-Borowski and Ricken 1997). Clones are positive for CD18, and they display activity of 3β-hydroxysteroid dehydrogenase and of alkaline phosphatase (Fig. 5.15a–f). Uptake of horseradish peroxidase speaks for the presence of macrophages (Fig. 5.15f). Progenitor cell markers such as Sox-2, Oct3/4, KIT, and CD14 are consistently present in clones (Fig. 5.15g, h). Progenitor cells probably represent the small cells of less than 10 μm in diameter at the ultrastructural level (Fig. 5.16). Collectively, angiogenesis, different cell death forms, leukocyte immigration, and the recruitment of progenitor cells fit into an acute inflammatory response being carefully mediated by the Myd88-related pathway of multipurpose INIM function (Takeda and Akira 2005; Takeuchi and Akira 2010; O'Neill and Bowie 2007).

5.1   Preovulatory Follicles and Superovulations in Species with a Short Estrous Cycle

5.1.1   Fibrinolytic Activity

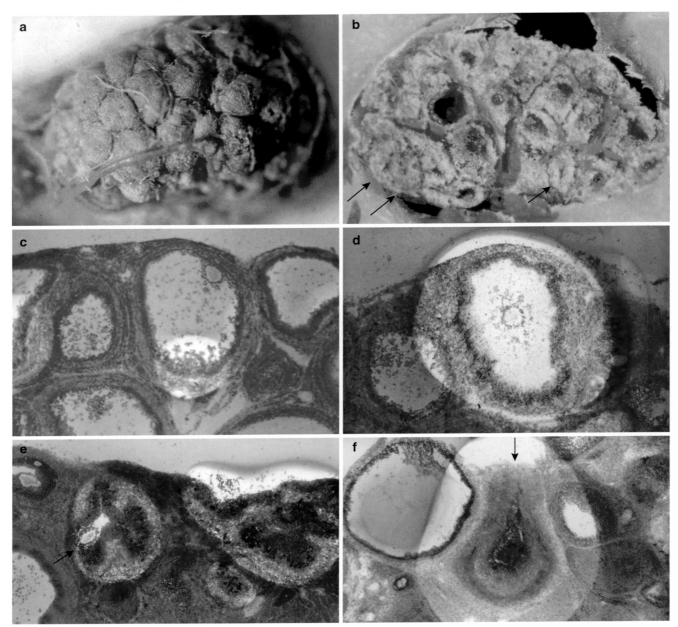

Fig. 5.1 Many preovulatory follicles develop after gonadotropic stimulation (s.c. injection of 50 IU PMSG at time "0" and 25 IU hCG/LH 60 h later) of 28-day-old golden hamsters. (**a, b**) The microvessel corrosion casts on day 4 after PMSG reveals nodules of follicles and of CLs (*arrows*) in the cross section in **b**. (**c–f**) On day 3 after PMSG application in H&E-stained cryostat sections, the fibrinolytic activity is depicted as lytic areas in a fibrin film. (**c, d**) Preovulatory follicles with detached cumulus oophorus and extended cumulus cells display fibrinolytic activity in a circumscribed area in **c** and in the whole area in **d**. (**e**) One of the two collapsed preovulatory follicles retains an oocyte (*arrow*). Fibrinolytic activity is at its maximum. (**f**) The developing CL with the rupture site apparent (*arrow*) has a strong fibrinolytic activity. x a,b:20; c,e:35; d,e,f:90 (Adapted from Spanel-Borowski and Heiss 1986; Spanel-Borowski et al. 1987)

5.1.2 Meiotic Division

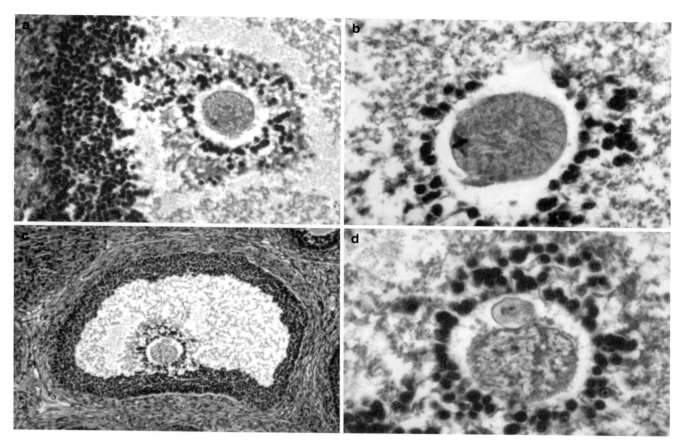

Fig. 5.2 First meiosis is resumed in preovulatory follicles of 27-day-old rat ovaries after s.c. injection of 30 IU PMSG at time "zero" and follicle rupture is inhibited by arginine vasotocin. A subcutaneous injection of 1 μg/0.1 was performed every 2 h for 24 h after PMSG.

(**a–d**) The first meiotic spindle (**a, b**) and the first polar body (**c, d**) are seen together with cumulus cell expansion in H&E-stained sections. x a:180; b,d.350; c:100 (Adapted from Spanel-Borowski et al. 1983)

5.1.3 Microvessel Corrosion Casts

5.1.3.1 Capillary Sprouts

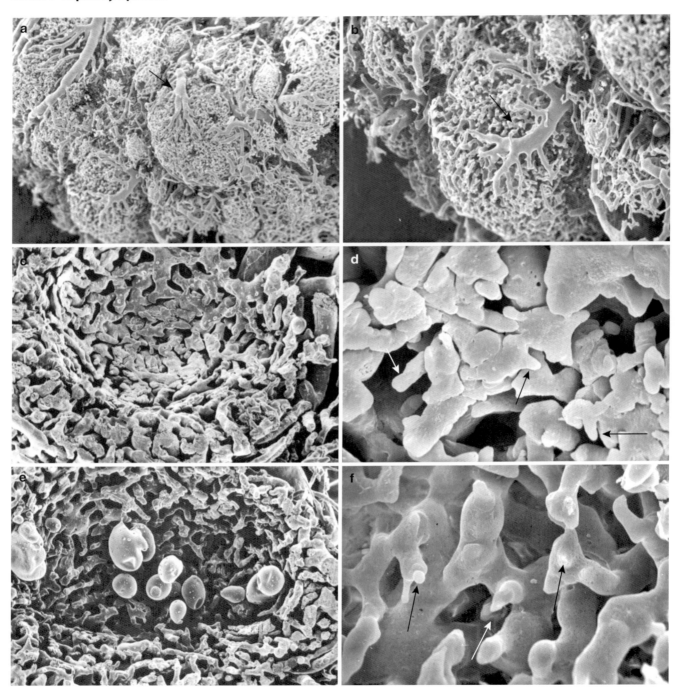

Fig. 5.3 Microvessels of preovulatory follicles from stimulated golden hamster ovaries (see Fig. 5.1) were casted with a polyester resin of low viscosity, and the three-dimensional structure of capillary sprouts was studied on day 4 after PMSG under the scanning electron microscope. (**a**, **b**) One major microvessel appears to support a preovulatory/mature follicle (*arrows*). The two thecal layers are difficult to see with the magnification of **b**. (**c**, **d**) The view of the capillary bed of the thecal layer reveals capillary sprouts concentrically aligned as blunt structures around the former basement membrane/antrum (*arrows*). The orientation of the sprouts is typical for follicle type I. (**e**, **f**) In follicle type II, capillary sprouts of the theca point radially toward the center of the former antrum. The orientation change indicates the disappearance of the basement membrane. Resin droplets in **e** show increased capillary permeability. x a:50; b:80; c,d:150; d.f:780 (Adapted from Spanel-Borowski et al. 1987)

5.1.3.2 Oocyte Sinusoid

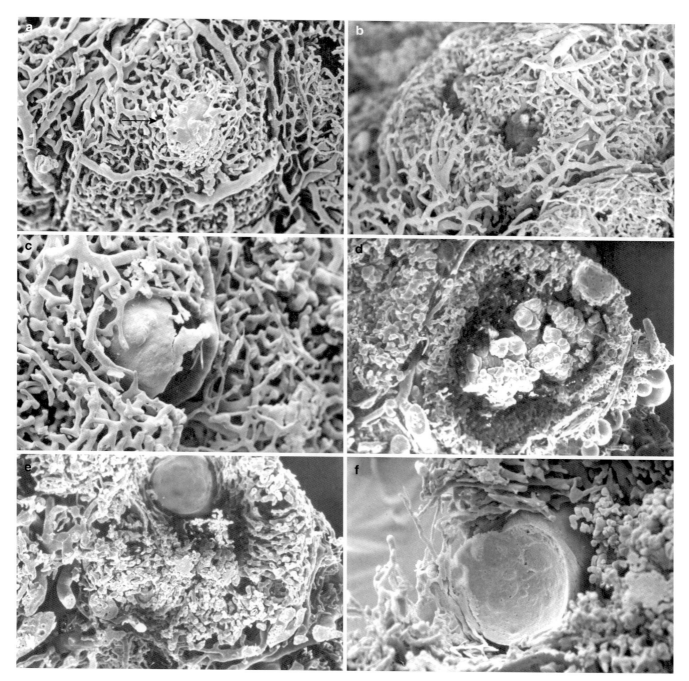

Fig. 5.4 The mature oocyte is embedded in a sinusoid of 130 μm^2 as deduced from microvessel corrosion casts of superovulated golden hamsters. Ovaries were viewed under the scanning electron microscope (see Fig. 5.1a, b). (**a**) The "resin lake" compares with a sinusoid at the follicle apex fed by a large microvessel (*arrow*). (**b**) The "punched-out circle" of 100 μm^2 is reminiscent of the rupture site after oocyte release. (**c**) The cap-like form is judged to represent an oocyte replica. The oocyte is entrapped in a resin lake caused by circumscribed capillary permeability. (**d**) The oocyte replica is generated during oocyte capture at the rupture site. Rupture is deduced from the radially orientated capillary sprouts and the accumulated resin droplets (Fig. 5.3e, f). (**e, f**) An oocyte replica is noted in a CL and could compare with unruptured follicle syndrome. x a,b,e:110; c.220; d:100; f:210 (Adapted from Löseke and Spanel-Borowski 1996)

5.1.3.3 Insufficient Microvascular Bed Caused by Repeated Superovulations

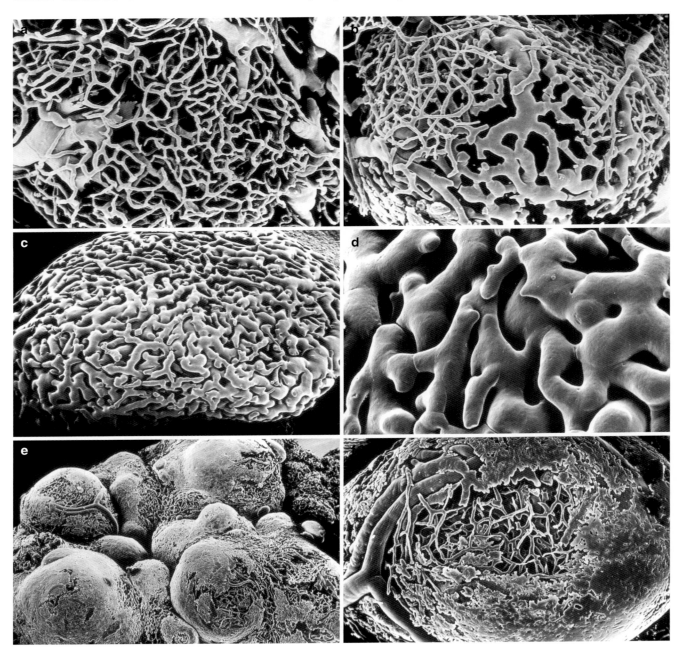

Fig. 5.5 Repeated superovulations are associated with an insufficiently developed microvascular bed of mature follicles examined under the scanning electron microscope. In 30-day-old golden hamsters, seven superovulations were induced with s.c. injections of 50 IU PMSG on day "0" and 25 IU hCG/LH at 60 h. Each subsequent treatment was 7 days apart, because the estrous cycle of superovulated golden hamsters lasts 7 days. The ovaries are depicted 6 h after the final LH injection in **a–d** and 36 h later in **e** and **f**. The defects are associated with hypo-ovulations shown in Fig. 7.12. (**a–d**) The microvascular bed of large antral follicles shows thread-like capillaries (**a**), two rarefied capillary layers (**b**), or heavily dilated capillaries (**c**). (**e, f**) The basket-like structure of the follicle wall disappears because of widespread resin leakages. x a:170; b:180; c:130; d:590; e:40; f:140 (Adapted from Löseke and Spanel-Borowski 1996)

5.2   Oocytes of Mouse Follicle Cultures After Treatment with Low-Density Lipoprotein (oxLDL) and Endoplasmic-Derived Multilamellar Bodies (EMBs)

5.2.1   Preantral Follicle Oocytes at 20 h of 100 µg/ml oxLDL Treatment

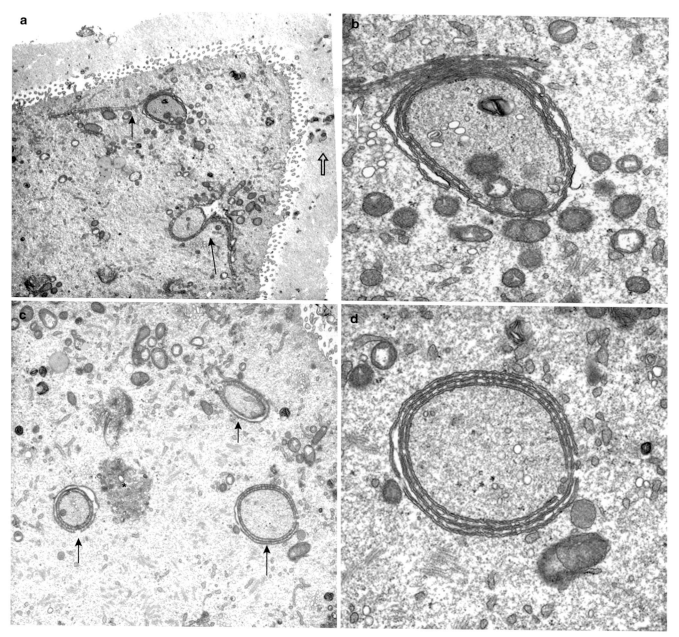

Fig. 5.6 Multilamellar bodies (EMBs) of type I are related with an organized tubular ER envelope in ultrathin sections. (**a**) The deformed oocyte possesses microvilli all along its oolemma, but only few trans-zonal projects are present (*open arrow*). The two developing EMBs appear to recruit tubules of ER into the envelope (*arrows*). (**b**) Magnification of the upper EMB in **a**. (**c, d**) Three forming EMBs (*arrows*) mark a triangle containing a multivesicular complex (*asterisk*). The upper EMB may represent an autophagosome because of its double membrane (*small arrow*), whereas the lower ones contain three membrane layers (*large arrows*) with some attached ribosome particles as shown in the magnified image in **d**. In the ooplasm, mitochondria with few cristae and with dense intermembrane space are present. x a:4'300; b:18'500, c:6'700, d:18'500 (Adapted from Spanel-Borowski et al. 2012)

5.2.2   Preantral Follicle Oocytes at 48 h of 100 μg/ml oxLDL Treatment

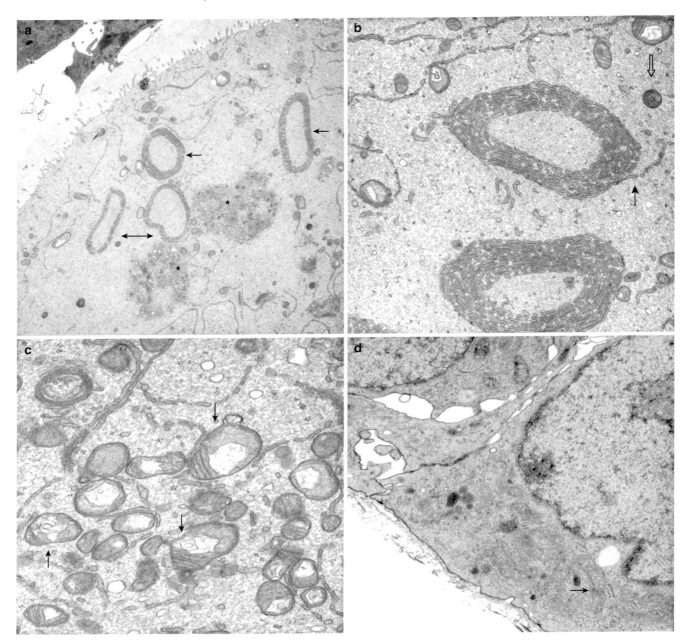

Fig. 5.7 Multilamellar bodies (EMBs) of type I depict a fully organized multitubular ER envelope in ultrathin sections as compared to those at 20 h of treatment. (**a**) Four EMBs (*arrow, double arrow*) and two multivesicular complexes (*asterisks*) are seen. (**b**) The envelope of the two EMBs consists of up to 12 layers of organized tubular ER. The upper one attracts a long tubule from the neighborhood (*arrow*). A degenerating mitochondrion with rings of circular cristae and dense matrix is depicted (*open arrow*). (**c, d**) Whereas mitochondria of the oocyte are frequently vacuolated with vesicular and decomposing membranes (*arrows*) indicating mild oxidative stress (**c**), mitochondria of granulosa cells remain intact with an elongated form and dense lamellae of the crista type (*arrow*; **d**). x a: 3'000; b,d:12'300; c:18'500 (Adapted from Spanel-Borowski et al. 2012)

5.2.3 Oocytes in Cumulus–Oocyte Complex Ovulated In Vitro 18 h After Stimulation by hCG/rEGF and Obtained from Preantral Follicles Cultured in the Presence of 100 µg/ml oxLDL for 13 Days

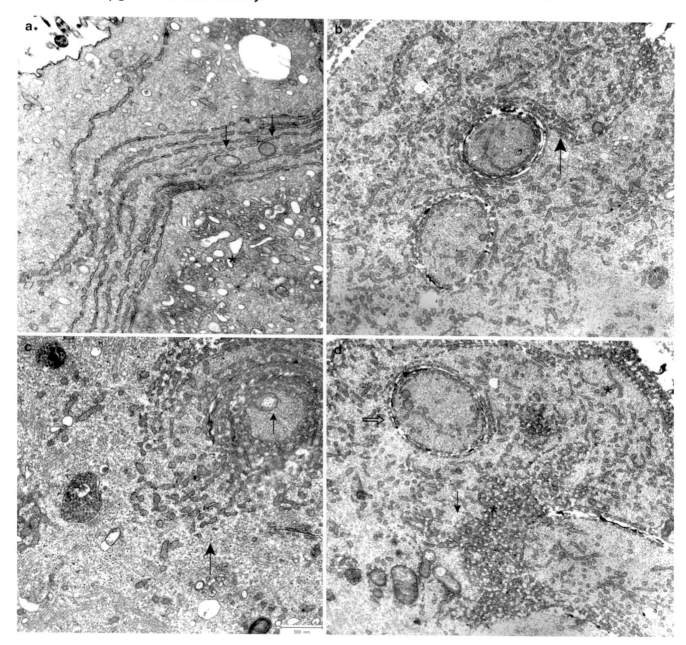

Fig. 5.8 Multilamellar bodies (EMBs) of type II display a vesicular smooth ER envelope in ultrathin sections. (**a**) The "tubules" near a multivesicular complex (*asterisk*) are composed of stacks of aligned vesicles enclosing a few degenerating mitochondria with single circular cristae or disrupted inner membranes (*arrows*). (**b, c**) The nonorganized/distributed vesicles belong to the branched form of smooth ER. The organized form builds the envelope of circular bodies, which embrace some degenerating mitochondria (*arrow* in **c**). There seems to be a conversion between the branched and organized form as depicted for two EMBs of type II (*long arrow* in **b** and **c**). (**d**) The ER displays the organized form in the envelope of the EMB (*open arrow*), whereas the branched form relates to distributed vesicles. They appear to be in connection with vesicle clusters (*arrow, asterisk*). x a:28'400; b:26'800; c:15'600, d:15'300 (Adapted from Spanel-Borowski et al. 2012)

5.3 Preovulatory Follicles from Species with a Long Ovarian Cycle

5.3.1 Intact and Freshly Ruptured Follicles

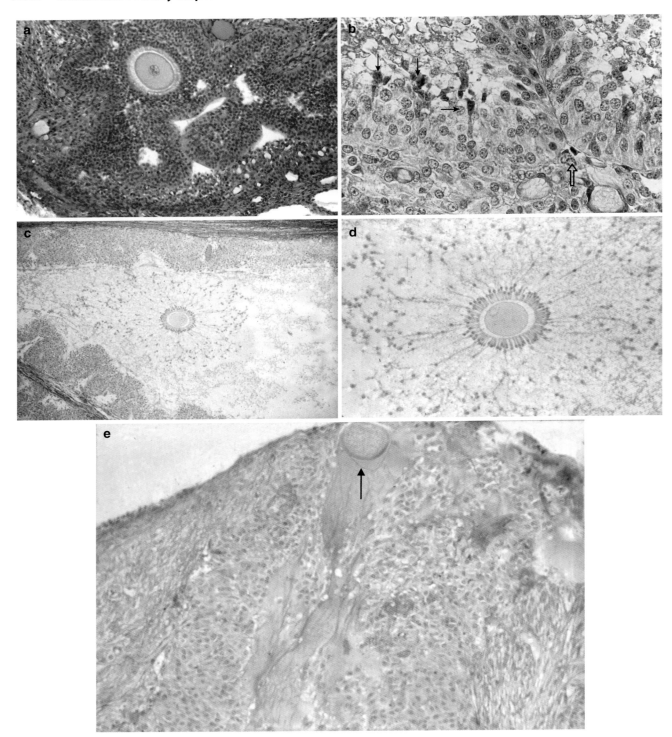

Fig. 5.9 Final differentiation of preovulatory follicles are shown for the dog (**a** and **b**) and the rabbit after ovulation induction (**c–e**). (**a, b**) The heavily gyrated follicle wall is collapsed enclosing the oocyte in diplotene arrest. The thecal cell layer infolds and starts to form the septum in the future CL (*open arrow in b*). The mural granulosa cell layer is infolded and contains some apoptotic cells close to the antrum (*arrows*). (**c, d**) While the mural granulosa is gyrated/infolded, the oocyte with expanded cumulus cells floats in the follicular fluid (**c**). The corona radiata cells close to the oocyte consists of cylindrical epithelial cells as a sign of mucification. (**d**) The oocyte is captured at the rupture site of a luteinizing follicle wall (*arrow*). H&E (**a, b**), HOPA (**c–e**) (Adapted from Spanel-Borowski et al. 1984). x a,d,e:100; b.430, c:40

5.3.2 Microvessels and Leukocytes In Situ Before Rupture

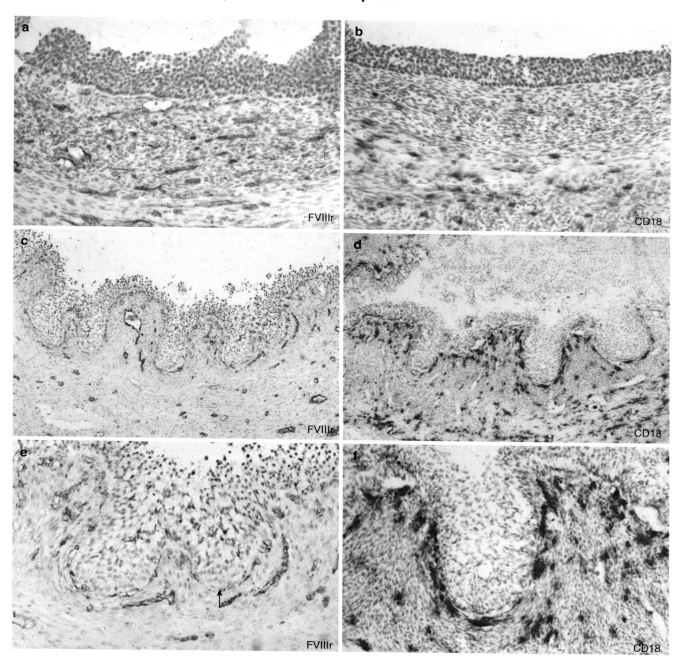

Fig. 5.10 Compared to antral follicles, preovulatory follicles of the bovine ovary develop a dense capillary network in the theca interna and recruit CD18-positive leukocytes. Immunostaining for FVIIIr antigen to detect endothelial cells (*left column*, **a, c, e**) and for CD18 as leukocyte marker (*right column*, **b, d, f**). (**a, b**) The thecal cell layer of antral follicles depicts uniformly distributed capillaries (**a**) and a few leukocytes (**b**). (**c–f**) In preovulatory follicles, infoldings of the theca interna show two layers of capillaries, one being associated with the basement membrane (**c** and **d**, *arrow* in **e**). Leukocytes densely populate the zone of the basement membrane between the granulosa and thecal cell layer in (**d** and **f**). x a,b:170; c,d:80; e:240; f:210 (Adapted from Spanel-Borowski et al. 1997)

5.3.3   Microvessels and Leukocytes In Situ After Rupture

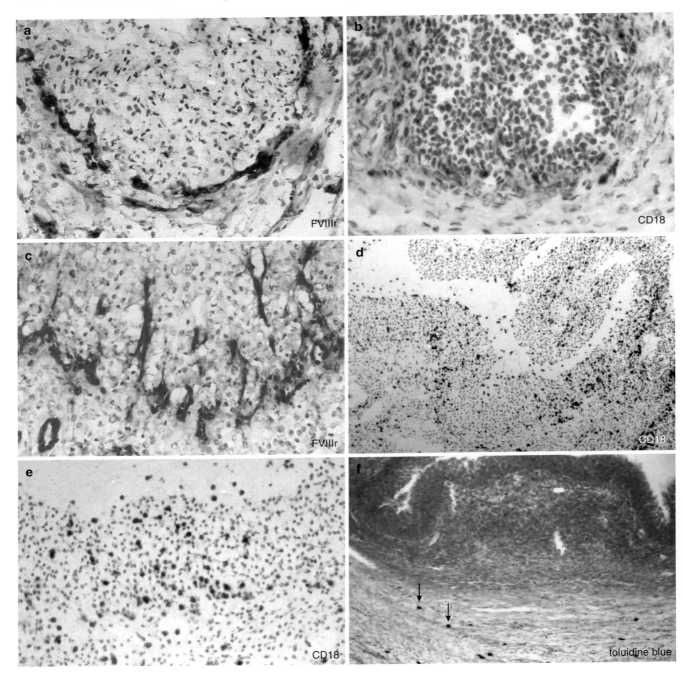

Fig. 5.11 Follicle rupture is accompanied by a change in capillary growth and in leukocyte invasion as seen in the bovine ovary. Sections are immunostained for FVIIIr antigen-positive endothelial cells (**a** and **c**), for CD18-positive leukocytes (**b, d, e**), and for mast cells with toluidine blue (**f**). (**a, c**) Capillaries are aligned concentrically around the basement membrane of the preovulatory follicle in (**a**). Capillaries vascularize the granulosa cell layer by pointing to the center of the antrum in **c**. (**b, d, e**) Leukocytes are absent in the granulosa cell layer of a preovulatory follicle in **b**. After rupture, the layer is heavily invaded by leukocytes (**d, e**). (**f**) The wall of a preovulatory follicle lacks mast cells, which are seen in the cortical stroma (*arrows*). x a,b,c:380; d:90; e:260; f:100 (Adapted from Spanel-Borowski et al. 1997; Reibiger and Spanel-Borowski 2000; Rohm et al. 2002)

5.3.4   Follicle Cell Harvests and Different Granulosa Cell Death Forms

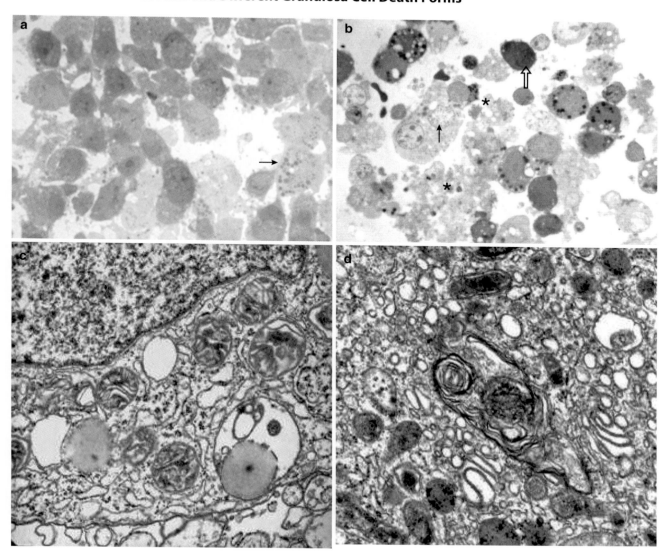

Fig. 5.12 A granulosa cell harvest from preovulatory follicles of women under IVF therapy is studied with semithin sections stained with toluidine blue in **a** and **b** and with ultrathin section in **c** and **d**. (**a**) Granulosa cells from normal-weight patients of younger reproductive age look intact and are rich in lipid droplets (*arrow*). (**b**) Granulosa cells from obese patients of older age show nuclei under lysis (*arrow*) and disintegrating cells (*asterisk*). Some cells with a condensed nucleus are hyperchromatic (*open arrow*). (**c**, **d**) Defective mitochondria with degenerative cristae in **c** might become a multilamellar body as shown in **d**. Both structures indicate oxidative stress. x a,b:350; c.37'000; d:24'900 (Adapted from Vilser et al. 2010)

5.3.5　Different Granulosa Cell Death Forms in Follicle Cell Harvests

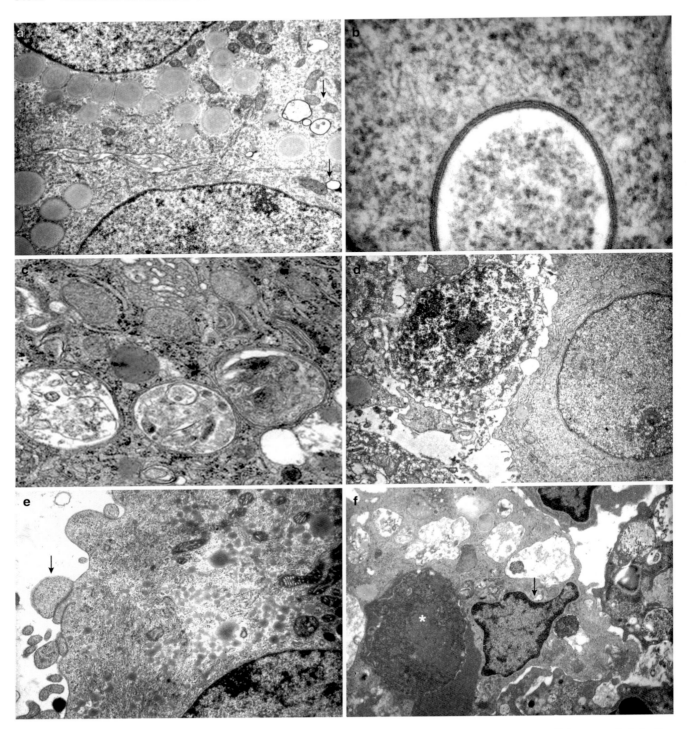

Fig. 5.13 Granulosa cells from human follicle harvests undergo different forms of cell death in ultrathin sections. (**a**) The two intact granulosa cells with a vesicular nucleus and many lipid droplets contain vesicle-like structures (*arrows*). (**b**) The "vesicles" represent annular junctions with a pentalaminar membrane and have originated by phagocytosis of tight junctions. (**c**) The three autophagosomes are recognized by the characteristic double membrane, which surrounds remnants of organelles. Autophagosomes indicate reparative autophagy that decreases during aging. (**d**) Lytic cell necrosis (*asterisk*) is seen to the left of an intact granulosa cell. (**e**, **f**) Blebbing of the cell membrane (*arrow*) is noted at the onset of apoptosis (**e**); the granulosa cell with condensed nuclear chromatin and many vacuoles represents a subsequent stage (*arrow* in **f**). The dark neighboring cell with a barely discernible nucleus (*asterisk*) could relate to an unusual nonapoptotic form. x a:13'600; b:51'100; c:41'600; d:8'600; e:14'100; f:5'300 (Adapted from Spanel-Borowski and Sterzik 1987; Vilser et al. 2010)

5.3.6   Cytokeratin-Positive Granulosa Cell Cultures

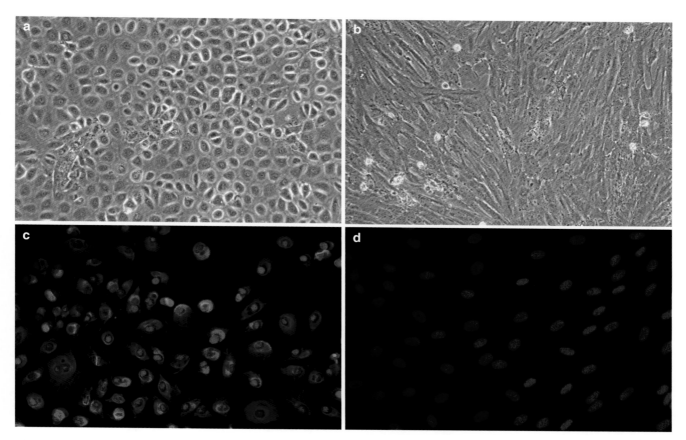

Fig. 5.14 Granulosa cell cultures with (**a** and **c**) and without (**b** and **d**) cytokeratin filaments were derived from human follicle aspirates. Cultures are imaged under the phase contrast microscope (*upper row*) or after staining with a pan cytokeratin antibody (*lower row*).

Cytokeratin-positive cells are candidates for dendritic-like cells, which regulate the expression of the toll-like 4 receptor under stimulation with oxLDL. x a,b:110; c,d:220 (Adapted from Serke et al. 2009, 2010)

5.3.7   Progenitor Cells In Vitro

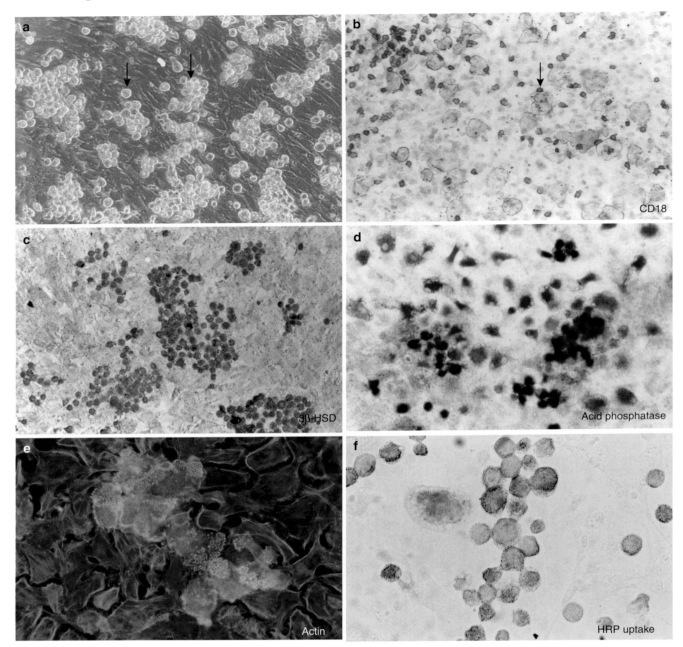

Fig. 5.15 Granulosa cell cultures derived from bovine antral follicles develop hematopoietic-like clones on day 10 of cultivation. (**a**) Clones adherent to the granulosa cell monolayer and isolated cells are noted under the phase microscope (*arrows*). (**b**) Immunostaining for CD18 reveals weakly stained cells, which are spread out on the monolayer like macrophages. Strongly stained cells have the size of lymphocytes (*arrow*) (**c**, **d**). Histochemical staining for activities of 3β-hydroxysteroid dehydrogenase (3β-HSD in **c**) and for acid phosphatase (**d**) shows a strong response of the cell clones. (**e**) Phalloidin staining of permeabilized cultures displays actin cables in granulosa cells and a granular- like actin cytoskeleton in the cluster cells. (**f**) Clone cells phagocytose horseradish peroxidase (HRP). Cultures were incubated with 400 μg peroxidase/ml for 60 min. Enzyme activity in brown was detected with the diaminobenzidine reaction. (**f**) Phalloidin staining of permeabilized cultures displays actin cables in cytokeratin-negative granulosa cells and a granular-like actin cytoskeleton in the cluster cells. (**g**, **h**) The double immunolocalization shows a clone with co-expression of CD14 and KIT, both markers of progenitor cells. x a,d:120; b:180; c.60; e:300; f:140; g,h:600 (Adapted from Spanel-Borowski and Ricken 1997; Merkwitz et al. 2010)

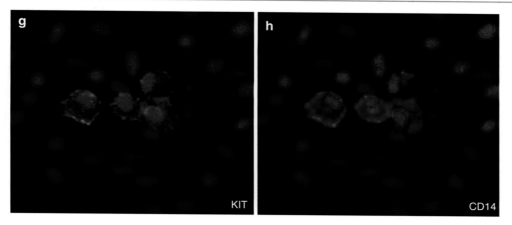

Fig. 5.15 (continued)

5.3.8   Progenitor Cells In Vitro and Ultrastructure

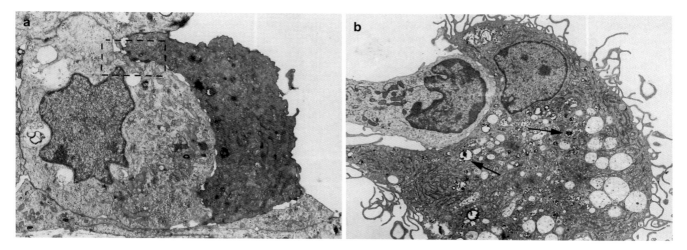

Fig. 5.16 The ultrastructure of hematopoietic-like clone cells is shown in 10-day-old granulosa cell cultures derived from bovine antral follicles. (**a**) Immature macrophage-like cells are connected by desmosomes (*rectangle*). (**b**) The large cell with abundant filopodia and granule-like structures is judged to be a mature macrophage. This is supported by the phagocytosis of horseradish peroxidase (*arrows* and Fig. 5.15f). The large cell embraces a small cell with a bean-shaped nucleus and without enzyme-positive granules. The cell might represent a precursor cell with the capacity to differentiate into endothelial cells. x a:7'900; b:5'000 (Adapted from Spanel-Borowski and Ricken 1997)

References

Bausenwein J, Serke H, Eberle K, Hirrlinger J, Jogschies P, Hmeidan FA, Blumenauer V, Spanel-Borowski K (2010) Elevated levels of oxidized low-density lipoprotein and of catalase activity in follicular fluid of obese women. Mol Hum Reprod 16:117–124

Bjersing L, Cajander S (1974) Ovulation and the mechanism of follicle rupture. VI. Ultrastructure of theca interna and the inner vascular network surrounding rabbit graafian follicles prior to induced ovulation. Cell Tissue Res 153:31–44

Byskov AG (1969) Ultrastructural studies on the preovulatory follicle in the mouse ovary. Z Zellforsch Mikrosk Anat 100:285–299

Chen Y, Klionsky D (2011) The regulation of autophagy – unanswered questions. J Cell Sci 124:161–170

de Bruin JP, Dorland M, Spek ER, Posthuma G, van Looman CW H, te Velde ER (2002) Ultrastructure of the resting ovarian follicle pool in healthy young women. Biol Reprod 66:1151–1160

Devine P, Perreault S, Luderer U (2012) Roles of reactive oxygen species and antioxidants in ovarian toxicity. Biol Reprod 86:27

Hanukoglu I (2006) Antioxidant protective mechanisms against reactive oxygen species (ROS) generated by mitochondrial P450 systems in steroidogenic cells. Drug Metab Rev 38:171–196

Honda A, Hirose M, Hara K, Matoba S, Inoue K, Miki H, Hiura H, Kanatsu-Shinohara M, Kanai Y, Kono T, Shinohara T, Ogura A (2007) Isolation, characterization, and in vitro and in vivo differentiation of putative thecal stem cells. Proc Natl Acad Sci USA 104:12389–12394

Hu Y, Betzendahl I, Cortvrindt R, Smitz J, Eichenlaub-Ritter U (2001) Effects of low O2 and ageing on spindles and chromosomes in mouse oocytes from pre-antral follicle culture. Hum Reprod 16:737–748

Inoue N, Matsuda F, Goto Y, Manabe N (2011) Role of cell-death ligand-receptor system of granulosa cells in selective follicular atresia in porcine ovary. J Reprod Dev 57:169–175

Kahnberg A, Enskog A, Brannstrom M, Lundin K, Bergh C (2009) Prediction of ovarian hyperstimulation syndrome in women undergoing in vitro fertilization. Acta Obstet Gynecol Scand 88:1373–1381

Kalesnikoff J, Galli SJ (2008) New developments in mast cell biology. Nat Immunol 9:1215–1223

Klionsky DJ (2007) Autophagy: from phenomenology to molecular understanding in less than a decade. Nat Rev Mol Cell Biol 8: 931–937

Klionsky DJ (2009) A work in progress. Autophagy 5:289

Kohchi C, Inagawa H, Nishizawa T, Soma G (2009) ROS and innate immunity. Anticancer Res 29:817–821

Korkhov VM (2009) GFP-LC3 labels organised smooth endoplasmic reticulum membranes independently of autophagy. J Cell Biochem 107:86–95

Korkhov VM, Milan-Lobo L, Zuber B, Farhan H, Schmid JA, Freissmuth M, Sitte HH (2008) Peptide-based interactions with calnexin target misassembled membrane proteins into endoplasmic reticulum-derived multilamellar bodies. J Mol Biol 378: 337–352

Löseke A, Spanel-Borowski K (1996) Simple or repeated induction of superovulation: a study on ovulation rates and microvessel corrosion casts in ovaries of golden hamsters. Ann Anat 178:5–14

Mehta JL, Chen J, Hermonat PL, Romeo F, Novelli G (2006) Lectin-like, oxidized low density lipoprotein receptor-1 (LOX-1). a critical player in the development of atherosclerosis and related disorders. Cardiovasc Res 69:36-45

Merkwitz C, Ricken AM, Lösche A, Sakurai M, Spanel-Borowski K (2010) Progenitor cells harvested from bovine follicles become endothelial cells. Differentiation 79:203–210

O'Neill LA, Bowie AG (2007) The family of five: TIR-domain-containing adaptors in toll-like receptor. Nat Rev Immunol 7:353–364

Qublan H, Amarin Z, Nawasreh M, Diab F, Malkawi S, Al-Ahmad N, Balawneh M (2006) Luteinized unruptured follicle syndrome: incidence and recurrence rate in infertile women with unexplained infertility undergoing intrauterine insemination. Hum Reprod 21:2110–2113

Reibiger I, Spanel-Borowski K (2000) Difference in localization of eosinophils and mast cells in the bovine ovary. J Reprod Fertil 118:243–249

Rohm F, Spanel-Borowski K, Eichler W, Aust G (2002) Correlation between expression of selectins and migration of eosinophils into the bovine ovary during the periovulatory period. Cell Tissue Res 309:313–322

Saller S, Merz-Lange J, Raffael S, Hecht S, Pavlik R, Thaler C, Berg D, Berg U, Kunz L, Mayerhofer A (2012) Norepinephrine, active norepinephrine transporter, and norepinephrine-metabolism Are involved in the generation of reactive oxygen species in human ovarian granulosa cells. Endocrinology 153:1472–1483

Serke H, Vilser C, Nowicki M, Hmeidan FA, Blumenauer V, Hummitzsch K, Lösche MT, Spanel-Borowski K (2009) Granulosa cell subtypes respond by autophagy or cell death to oxLDL-dependent activation of the oxidized lipoprotein receptor 1 and toll-like 4 receptor. Autophagy 5:991–1003

Serke H, Bausenwein J, Hirrlinger J, Nowicki M, Vilser C, Jogschies P, Hmeidan FA, Blumenauer V, Spanel-Borowski K (2010) Granulosa cell subtypes vary in response to oxidized low-density lipoprotein as regards specific lipoprotein receptors and antioxidant enzyme activity. J Clin Endocrinol Metab 95:3480–3490

Snapp EL, Hegde RS, Francolini M, Lombardo F, Colombo S, Pedrazzini E, Borgese N, Lippincott-Schwartz J (2003) Formation of stacked ER cisternae by low affinity protein interactions. J Cell Biol 163:257–269

Spanel-Borowski K (2011) Footmarks of innate immunity in the ovary and cytokeratin-positive cells as potential dendritic cells, vol 209, Advances in anatomy, embryology and cell biology. Springer, Heidelberg, ISBN 978-3-642-16076-9

Spanel-Borowski K (2011) Ovulation as danger signaling event of innate immunity. Mol Cell Endocrinol 333:1–7

Spanel-Borowski K, Sterzik K (1987) Ultrastructure of human preovulatory granulosa cells in follicular fluid aspirates. Arch Gynecol 240:137–146

Spanel-Borowski K, Heiss C (1986) Luteolysis and thrombus formation in ovaries of immature superstimulated golden hamsters. Aust J Biol Sci 39:407–416

Spanel-Borowski K, Ricken AM (1997) Evidence for the maintenance of macrophage-like cells in long-term bovine granulosa cell cultures. Cell Tissue Res 288:529–538

Spanel-Borowski K, Vaughan LY, Johnson LY, Reiter RJ (1983) Increase of intra-ovarian oocyte release in PMSG-primed immature rats and its inhibition by arginine vasotocin. Biomed Res 4:71–82

Spanel-Borowski K, Thor-Wiedemann S, Pilgrim C (1984) Cell proliferation in the dog (beagle) ovary during proestrus and early estrus. Acta Anat 118:153–158

Spanel-Borowski K, Amselgruber W, Sinowatz F (1987) Capillary sprouts in ovaries of immature superstimulated golden hamsters: a SEM study of microcorrosion casts. Anat Embryol 176:387–391

Spanel-Borowski K, Rahner P, Ricken AM (1997) Immunolocalization of CD18-positive cells in the bovine ovary. J Reprod Fertil 111:197–205

Spanel-Borowski K, Sass K, Löffler S, Brylla E, Sakurai M, Ricken AM (2007) KIT receptor-positive cells in the bovine corpus luteum are primarily theca-derived small luteal cells. Reproduction 134:625–634

Spanel-Borowski K, Nowicki M, Borlak J, Trapphoff T, Eichenlaub-Ritter U (2012) Endoplasmic reticulum-derived multilamellar bodies in oocytes of mouse follicle cultures under oxidised low density lipoprotein. Cells Tiss Org Epub ahead in print. DOI.10.1159/000340039

Sun F, Betzendahl I, Pacchierotti F, Ranaldi R, Smitz J, Cortvrindt R, Eichenlaub-Ritter U (2005) Aneuploidy in mouse metaphase II oocytes exposed in vivo and in vitro in preantral follicle culture to nocodazole. Mutagenesis 20:65–75

Szotek P, Pieretti-Vanmarcke R, Masiakos P, Dinulescu D, Connolly D, Foster R, Dombkowski D, Preffer F, Maclaughlin D, Donahoe P (2006) Ovarian cancer side population defines cells with stem cell-like characteristics and mullerian inhibiting substance responsiveness. Proc Natl Acad Sci USA 103:11154–11159

Takeda K, Akira S (2005) Toll-like receptors in innate immunity. Int Immunol 17:1–14

Takeuchi O, Akira S (2010) Pattern recognition receptors and inflammation. Cell 140:805–820

Tatemoto H, Muto N, Sunagawa I, Shinjo A, Nakada T (2004) Protection of porcine oocytes against cell damage caused by oxidative stress during in vitro maturation: role of superoxide dismutase activity in porcine follicular fluid. Biol Reprod 71:1150–1157

Tatone C, Amicarelli F, Carbone MC, Monteleone P, Caserta D, Marci R, Artini PG, Piomboni P, Focarelli R (2008) Cellular and molecular aspects of ovarian follicle ageing. Hum Reprod Update 14:131–142

Tilly JL (1996) Apoptosis and ovarian function. Rev Reprod 1:162–172

Tilly J, Rueda B (2008) Minireview: stem cell contribution to ovarian development, function, and disease. Endocrinology 149:4307–4311

Tsukamoto S, Kuma A, Mizushima N (2008) The role of autophagy during the oocyte-to-embryo transition. Autophagy 4:107–1078

van Wezel IL, Dharmarajan AM, Lavranos TC, Rodgers RJ (1999) Evidence for alternative pathways of granulosa cell death in healthy and slightly atretic bovine aantral follicles. Endocrinology 140:2602–2612

Vilser C, Hueller H, Nowicki M, Hmeidan FA, Blumenauer V, Spanel-Borowski K (2010) The variable expression of lectin-like oxidized low-density lipoprotein receptor (LOX-1) and signs of autophagy and apoptosis in freshly harvested human granulosa cells depend on gonadotropin dose, age, and body weight. Fertil Steril 93:2706–2715

Whitaker BD, Knight JW (2008) Mechanisms of oxidative stress in porcine oocytes and the role of anti-oxidants. Reprod Fertil Dev 20:694–702

Corpus Luteum Life Cycle with Focus on Capillary Sprouting and Regression, Eosinophils, and KIT-Positive Cells

The wall of the ruptured preovulatory follicle transforms into the CL. The ephemeral endocrine gland is classified into stages of development, secretion, and regression. The endocrine function fades with each ovarian cycle (Niswender et al. 2000; Stocco et al. 2007). The CL compares with a benign tumor, which is finally removed without or with some residues like the corpus albicans in women (Devoto et al. 2009). The cyclic coming and going of CLs represents a high turnover rate that is perfect in controlling tissue homeostasis during the reproductive period. Dismissal of the CL starts with functional luteolysis that is characterized by an acute stop in progesterone secretion. Structural luteolysis indicates degeneration and cellular changes in the CL architecture. Depending on the species, structural luteolysis is acute or extended. The acute process occurs in golden hamsters and is striking after superovulations (Gaytán et al. 2001; Spanel-Borowski and Heiss 1986). Superovulations in immature golden hamsters induced by pregnant mare serum gonadotropin (PMSG) on day 0 and by LH 60 h later extend the 4-day-ovarian cycle to a 7-day cycle. Within 4 days after ovulation, the CL life cycle progresses with the presence and subsequent absence of capillaries, which are replaced by apoptotic bodies. In fact, the previous CLs are undetectable at the onset of the next ovarian cycle on day 8 after PMSG application. For this reason, resin filling of the microvascular bed is possible on day 4 after PMSG application, but not on day 7 (Fig. 6.1). Laminin immunostaining for basement membrane localization reveals a fully developed microvascular bed only on day 4 after PMSG application (Fig. 6.2a–d). Subsequently, abundant apoptotic bodies enclosed by a disintegrating basement membrane indicate the death of endothelial cells on day 6 (Fig. 6.2e, f). The striking apoptotic event and the degenerated steroidogenic luteal cells explain the rapid decrease in weight from roughly 120 to 40 mg between days 5 and 7 after PMSG application (Fig. 6.2 g, h) (Spanel-Borowski and Heiss 1986).

The microvascular bed is judged as a main player in the CL life cycle (Niswender et al. 2000). The bed develops after follicle rupture when capillaries sprout from the theca into the previous avascular granulosa cell layer. Then capillaries mature into an organized microvascular bed. Its death terminates the CL life cycle. This complex process requires a tight interaction between angiogenic and endothelial cell molecules. Expression and function of the VEGF-system, the endothelin-system, the angiopoietin-system, and the RAS system are found for mature follicle and for CLs (Kaczmarek et al. 2005; Klipper et al. 2010; Nishigaki et al. 2011; Gonçalves et al. 2012). The sequence from capillary sprouting to opening of the vascular lumen is beautifully demonstrated in CLs from superovulated hamsters at the ultrastructural level. The enzyme HRP, which was intravenously injected at 10 mg/100 g body weight just before the hamster's death, permeates endothelial cells. The HRP activity is detected at the basement membrane by histochemical reaction thus localizing young sprouts (Spanel-Borowski and Mayerhofer 1987). Early sprouts on day 4 after PMSG application demonstrate a high endothelial cell permeability as is deduced from a basement membrane with strong HRP positivity (Fig. 6.3). This advantage helps to detect sprouts resembling a precursor cell (Fig. 6.3a, b). Sprouting starts in the mother capillary; it releases activated endothelial cells through a gap of the basement membrane (Fig. 6.3a, c). The HRP-positivity in the gap of sprout tips indicates the future vascular lumen (Fig. 6.3d–f). Opening of the gap is likely due to changing physical forces in the mother capillary. Advanced sprouts on day 5 after PMSG treatment align in endothelial cell bands delineated by low HRP-positivity both at the basement membrane and in the arising vascular lumen (Fig. 6.4). Sprouts developing into capillaries are captured in the process of decay on day 6 after PMSG application. All capillaries are appointed to die immediately, more through apoptosis than necrosis (Fig. 6.5a–d). Rapid elimination of cellular debris obviously functions by way of still-intact vessels (Fig. 6.5e, f). The event compares with vascular luteolysis and explains the ability of golden hamster ovaries to remove CLs within one ovarian cycle (Chap. 8).

In rat and mouse ovaries as well as in ovaries from cows and women, the CL involution runs over several ovarian

K. Spanel-Borowski, *Atlas of the Mammalian Ovary*,
DOI 10.1007/978-3-642-30535-1_6, © Springer-Verlag Berlin Heidelberg 2012

cycles and represents an extended luteolytic process. Thus, CLs from species with a 4-day estrous cycle (mice, rats) seem to be subjected to a similar luteolytic mechanism as CLs evolving during a 21- and 28-day cycle (cows, women), respectively. A large CL, like that found in cows, is a valuable model for isolating angiogenic factors such as VEGF, angiopoietin-2, and endothelin-1 and their receptors depending on different functional stages (Schams and Berisha 2004; Meidan and Levy 2007). The bovine CL compares with its human counterparts in size and also in roughly the same number of days needed for the various stages of development, secretion, and regression (Devoto et al. 2009) (Fig. 6.6a). Bovine CLs are easily available at slaughterhouses, and their functional stages can be roughly defined at the macroscopical level. For ethical reasons, fresh CLs of women are seldom obtained after surgical interventions. Paraffin blocks with samples of CLs can be obtained from the archives of various institutes of pathology. The early stage of CL development displays gyration of the parenchyma, which follows the gyrated preovulatory follicle wall. In the CL periphery, small thecal lutein cells infold the layer of granulosa lutein cells (Fig. 6.6b, c). Small lutein cells form an incomplete septum giving the CL a pseudolobular architecture. The width of the theca-derived septum decreases at the secretory stage. It is interesting that mast cells never populate the CL, a finding that is in line with the absence of mast cells in the preovulatory follicle wall (compare Figs. 5.11f and 6.6d). In terms of immunosurveillance, mast cells seem to be excluded from CL life regulation.

The macroscopical classification of the CL stages in the cow is confirmed at the light microscopical level (Spanel-Borowski et al. 1997). Leukocyte distribution and the microvascular bed undergo characteristic changes (Fig. 6.7). These changes seem to depend on each other. As long as sprouting continues, CD18-positive leukocytes are preferentially limited to the septum area. Uniform leukocyte presence coincides with the developed microvascular bed at the secretory stage (Fig. 6.7a–d). Leukocytes increase in density at the onset of regression and are clustered during advanced regression between prominent arterioles that look like shunt vessels (Fig. 6.7e–h). Their origin is related to young arterioles containing a discrete layer of proliferating smooth muscle cells already at the stage of secretion (Bauer et al. 2003). The media layer conspicuously thickens and undergoes fibroelastosis similar to the vascular fibroelastosis seen during chronic hypertension. As is concluded for the bovine and human CL, capillaries disappear during luteolysis, but arterioles survive. The inflammatory-like response in the early CL occurs obviously in continuity with the inflammatory response in the preovulatory follicle. Yet changes in leukocyte subtype are evident. At early CL development, 90 % of the CD18-positive pool comprises eosinophils, which densely populate the luteinizing thecal cell layer (Reibiger and Spanel-

Borowski 2000). Eosinophils become sparse in the subsequent stages (Fig. 6.8a–d). The healing phase of the ruptured follicle compares with early CL formation. Here, immunocompetent eosinophils could play an anti-inflammatory role relating to connective tissue remodeling and angiogenesis (Munitz and Levi-Schaffer 2004; Nissim Ben Efraim and Levi-Schaffer 2008). Eosinophils are replaced by increasing densities of T cells, in particular CD8-positive/cytotoxic cells, and of macrophages, which are attracted through the monocyte-chemoattractant protein 1 and undergo local proliferation at the advanced stage of regression in the bovine CL (Bauer et al. 2001). Concerning eosinophil recruitment at early stages, a tachykinin with similarities to substance P (SP) in immunostaining might be influential (Reibiger et al. 2001; Debeljuk 2006). Substance P exerts inflammatory effects through the neurokinin-1 receptor by the release of inflammatory mediators (Severini et al. 2002). The SP-like response, which is depicted as a fiber-like network either within or between steroidogenic luteal cells in immunostaining, is limited to the periphery of the developing CL. The network disintegrates in the subsequent stages (Fig. 6.8e–h), which are associated with a minimum of eosinophils. Although strong evidence is given for the presence of the tachykinin–tachykinin receptor system in the bovine CL at the mRNA and protein level, and the influence of tachykinins on ovarian functions has been reported (Reibiger et al. 2001; Brylla et al. 2005; Debeljuk 2006), the final proof of its existence is still missing. The reason is that preincubation of the SP antiserum with the SP peptide does not inhibit the network-like immunoresponse in the developing CL (Spanel-Borowski 2011). Provided that the SP antiserum contains an antibody against an unknown tachykinin member, absorption with the SP peptide could be the cause for the pitfall encountered in control immunostainings.

The KIT receptor controls proliferation, differentiation, and migration of cells (Rönnstrand 2004). KIT expression characterizes bone-marrow-derived progenitor cells either of hematopoietic or of endothelial/epithelial cell lineage. KIT plays a key role in the growth and maturation of follicles as well as in the CL life cycle (Spanel-Borowski et al. 2007). In the bovine CL of early developmental stage, thecal lutein cells mount a band-like KIT-positive immunoresponse (Fig. 6.9a, b). The KIT-positive cells intermingle with granulosa lutein cells in the CL at advanced developmental stages (Fig. 6.9c, d). A KIT-positive cell network appears in the outer zone of the CL, showing substantial expansion at the stage of secretion (Fig. 6.9e–h). Colocalization with steroidogenic enzymes defines KIT-positive cells as small luteal cells (Spanel-Borowski et al. 2007), which represent the majority. A minority of cells are defined as leukocytes with KIT expression, as is revealed by double immunofluorescence localization for KIT and CD45 in the CL at early developmental stages (Fig. 6.10).

Collectively, the CL of early developmental stage is judged as a healing period after physiological wounding of the follicle wall. The control tower appears to be the previous thecal cell layer guiding the recruitment of eosinophils and some progenitor cells as well as the up-regulation of KIT-positive steroidogenic thecal cells. Eosinophils belong to the specific immunocompetent cells of INIM like mast cells (Turvey and Broide 2010; Medzhitov 2010). Eosinophils could mediate angiogenesis and connective tissue remodeling through activation of angiogenic factors and of transforming growth factor β (Munitz and Levi-Schaffer 2004). The up-regulation of KIT in steroidogenic thecal cells might cause their migration from the theca between the luteinizing granulosa. Thus, small KIT-positive theca lutein cells mobilize and interact with large CK-positive granulosa lutein cells. Provided that the CK-positive granulosa cells are finally accepted as a novel form of dendritic cell (Chap. 10), functions of CK-positive cells could be modified by the SP-like tachykinin as shown for classic dendritic cells (Dunzendorfer and Wiedermann 2001). A change in the function of CK-positive cells might mediate luteolysis. It is an acute process in golden hamster CLs expressing signs of acute inflammation together with striking apoptosis. Extended luteolysis in bovine CLs represents a chronic inflammatory response with rare apoptotic bodies. They appear to be of less importance than cell-death-related autophagy (Gaytán et al. 2001, 2008; Choi et al. 2011).

6.1   Acute Structural Luteolysis in Superovulated Golden Hamster Ovaries

Golden hamsters develop a short ovarian/estrous cycle and acute luteolysis. Hamsters aged 29 and 30 days were s.c. injected with 50 IU PMSG at time "0" and 25 IU hCG/LH 60 h later. The ovarian stimulation extends the normal 4-day estrous cycle to 7 days. A precise sequence of CL changes is seen at the light and ultrastructural levels between days 4 and 7 after PMSG application. For Figs. 6.3, 6.4, 6.5, horseradish peroxidase (HRP) was used as an endothelial cell tracer to detect capillary sprouts. The jugular vein of anesthetized hamsters was injected with 1 % HRP solution and the left ovary removed 1 min later. The enzyme was localized by ultrahistochemistry. HRP activity was highly positive on day 4 after PMSG, decreased thereafter, and was negative on PMSG day 7

6.1.1   Microvessel Corrosion Casts

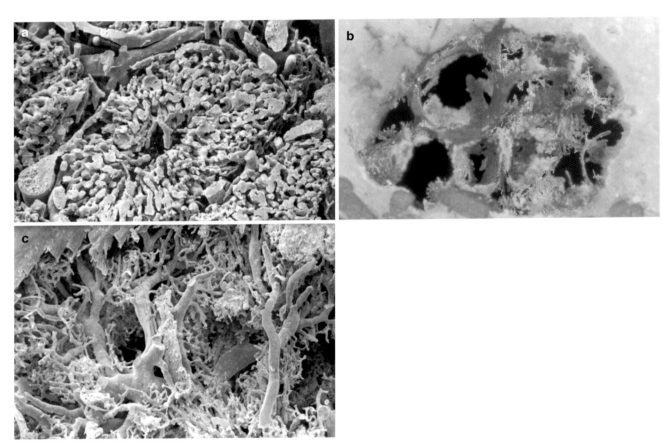

Fig. 6.1 Microvessels of CLs from superovulated immature golden hamsters (see Figs. 5.1a, b and 5.3) were casted with polyester resin and investigated under the scanning electron microscope. (**a**) The radial growth of capillary sprouts seen on day 4 after PMSG application is reminiscent of a wheel's spokes. (**b, c**) On day 6 after PMSG application, capillary-free areas are noted as defects at the macroscopical level in **b**. They are limited by larger microvessels under higher magnification in **c**. Defects correspond to regressed CLs in advanced stage. x a.b:110; c:40 (Adapted from Spanel-Borowski et al. 1987)

6.1.2   Corpus Luteum Stages and the Microvascular Bed: Histology

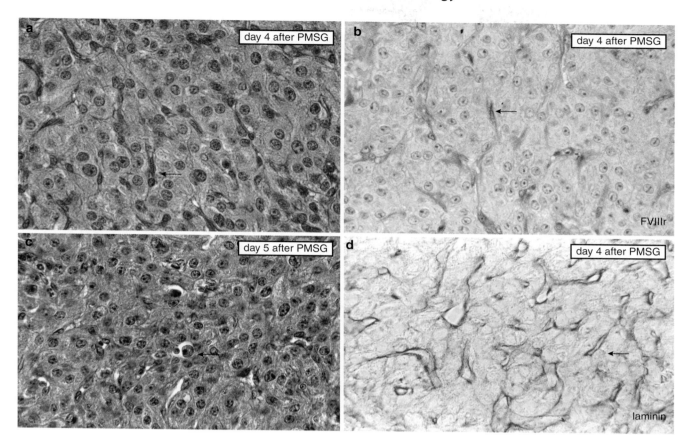

Fig. 6.2 The CL life cycle is shown for ovaries of superovulated imma-ture golden hamsters. Sections were stained with H&E to show changes between days 4 and 7 after PMSG application (*left column*) and immu-nostained for FVIIIr-positive endothelial cells or for laminin to detect the basement membrane (*right column*). (**a, b, d**) On day 4 after PMSG application, the CL of secretory stage depicts steroidogenic luteal cells and FVIIIr-positive endothelial cells (*arrow*). The laminin staining in **d** is superior in revealing the developed microvascular bed than the FVIIIr staining in **b**. (**c**) On day 5 after PMSG, a single apoptotic cell (*arrow*) and the disappearance of endothelial cells signify the onset of luteolysis. (**e, f**) On day 6 after PMSG, the regressing CL contains many apoptotic bodies enclosed by the basement membrane of previous capillaries. (**g, h**) On day 7 after PMSG application, most apoptotic bodies have disap-peared. Luteal cells show fatty degeneration. Fragments of the basement membrane are clustered. x a,b,c,d,e,g:240; f:360; h:120 (Adapted from Spanel-Borowski and Heiss 1986)

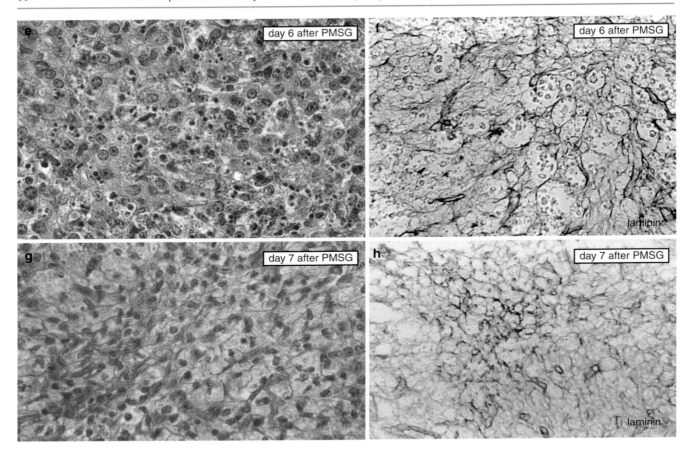

Fig. 6.2 (continued)

6.1.3   Ultrastructure of Capillary Sprouts

6.1.3.1   Early Formation

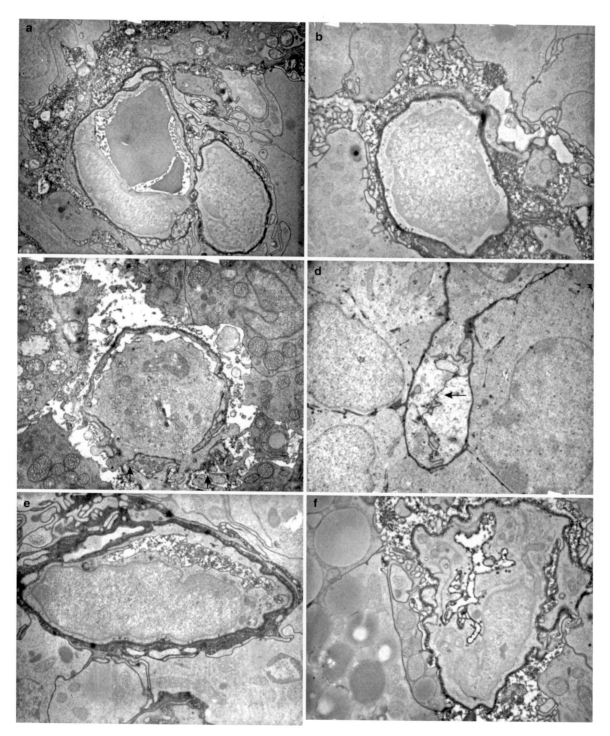

Fig. 6.3 Early formation of capillary sprouts is seen in the CL of immature golden hamsters on day 4 after PMSG application. Horseradish peroxidase (HRP) activity is found at the basement membrane, in micropinocytotic vesicles, and in the intercellular space. (**a**, **b**) The sprout with a large nucleus and with sparse cytoplasm is reminiscent of a precursor cell. (**c**) A high endothelial cell forms pseudopods (*arrows*) to emigrate through the basement membrane of the mother capillary with a flat endothelium. (**d**) The tip of a capillary sprout shows a highly convoluted HRP-positive gap (*arrow*) in between active luteal cells. The gap likely represents the meeting-point of two pseudopods. (**e**, **f**) In both cross-cuts through early capillary sprouts, HRP activity is noted in the arising vascular lumen and in the perivascular space alluding to high vascular permeability. x a: 6'000; b: 8'700; c:7'900; d:4'700; e,f:9'000 (Adapted from Spanel-Borowski and Mayerhofer 1987)

6.1.3.2 Advanced Formation

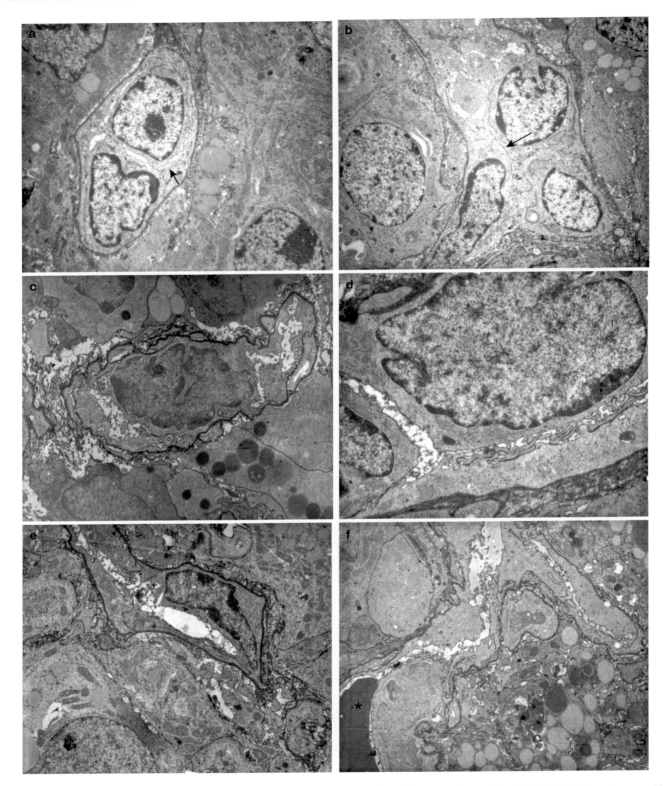

Fig. 6.4 Advanced formation of capillary sprouts is seen in the CL of immature golden hamsters on day 5 after PMSG application. Compared to Fig. 6.3, HRP activity has decreased as a sign of decreased endothelial cell permeability. (**a, b**) Sprouts consist of high endothelial cells as seen in the cross-cut and longitudinal cut. The interendothelial space is less pronounced than in Fig. 6.3d (*arrow*). (**c, d**) The interendothelial space of capillary sprouts is opening. The luminal cell membranes with a weak HRP activity are either adjacent or interdigitate. (**e, f**) The interendothelial space is dilated at a branching area where an erythrocyte squeezes in (**f**, *asterisk*). x a:5'400; b:3'600; c:7'500;d:8'300; e:3'500; f:2'900 (Adapted from Spanel-Borowski and Mayerhofer 1987)

6.1.3.3 Regressing Sprouts

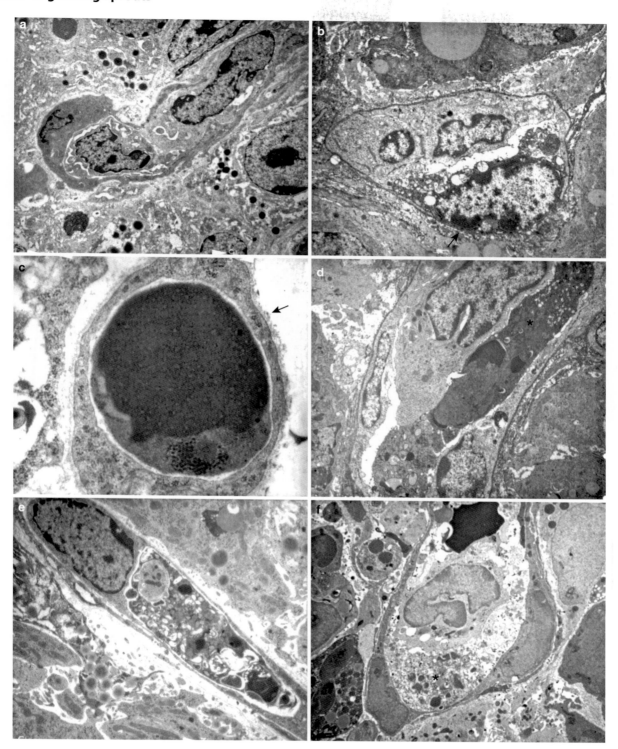

Fig. 6.5 Regressing capillary sprouts in the CL of immature golden hamsters show endothelial cells under necrosis and apoptosis on day 6 after PMSG application. Compared to Figs. 6.3 and 6.4, HRP activity is negligible. (**a**) The capillary sprout is degenerating according to the shrunken nuclei and the electron-dense cytoplasm. (**b**) The endothelial cell of a cross-cut sprout undergoes necrosis and appears to lack the basement membrane (*arrow*). The disintegrating cytoplasm is diffusely penetrated by HRP. (**c**) A capillary sprout with the basement membrane (*arrow*) surrounds an apoptotic body with highly condensed chromatin. (**d**) The longitudinal cut of a regressing capillary sprout depicts a classic apoptotic cell with marginal chromatin condensation and partially maintained organelles (*asterisk*) in the center. Intact endothelial cells are seen in the periphery. (**e**) The large sprout contains an advanced stage of cell destruction. (**f**) The venule contains a lot of cellular debris (*asterisk*) in addition to a monocyte and an erythrocyte x a:3'500; b:5'000; c:22'200;d:5'200; e:4'600; f:3'000 (Adapted from Spanel-Borowski and Mayerhofer 1987)

6.2   Bovine Ovaries with Extended Structural Luteolysis

6.2.1   Overview of Corpus Luteum Stages

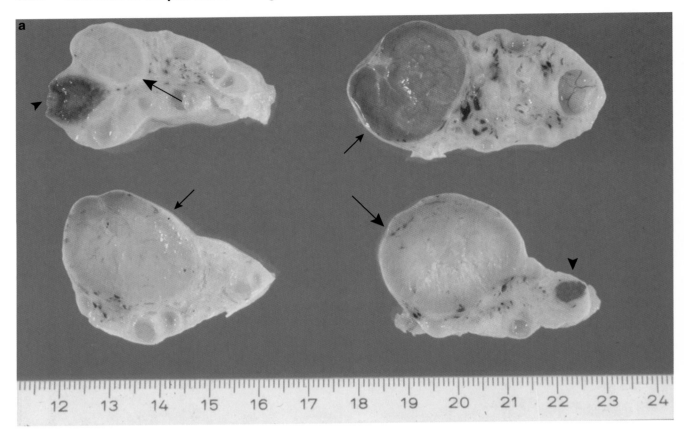

Fig. 6.6 Different functional stages of the CL are shown for the cow (**a** and **d**) and for the human (**b** and **c**). (**a**) The CL of early developmental stage is red-brown and small at early stages (*arrowhead*). The soft tissue of the CL in the secretory stage becomes pale brown (*small arrows*), and the CL in the stage of regression turns to yellow in addition to increasing tissue firmness (*long arrows*). In the very advanced stage of regression, the CL is small, firm, and orange (*arrowhead*). (**b, c**) The CL of early developmental stage consists of highly gyrated luteal tissue in the van-Gieson-stained section. Gyration has started in the former preovulatory follicle wall (see Figs. 5.9 and 5.10). The periphery of the CL shows thecal lutein cells, which form the septum (*arrow*) in between the granulosa lutein cells (**c**). (**d**) Mast cells (*arrows*) are absent in the fully developed CL in the toluidine blue-stained section. This is similar to the absence of mast cells in the mature follicle (Fig. 5.11f). x a: scale; b:3; c:240; d:30 (Adapted from Spanel-Borowski 2010)

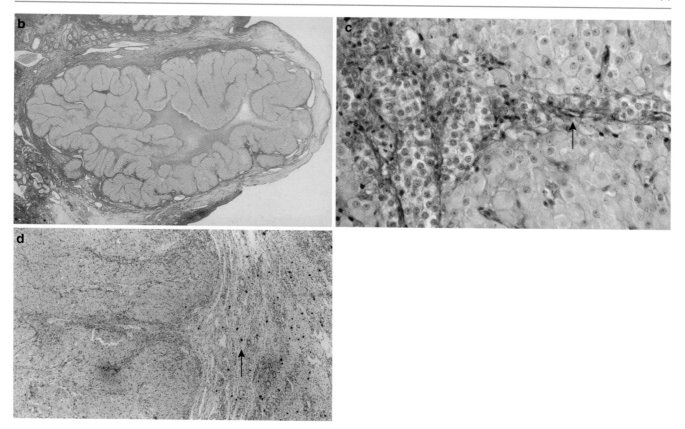

Fig. 6.6 (continued)

6.2.2   Leukocytes and Microvessels

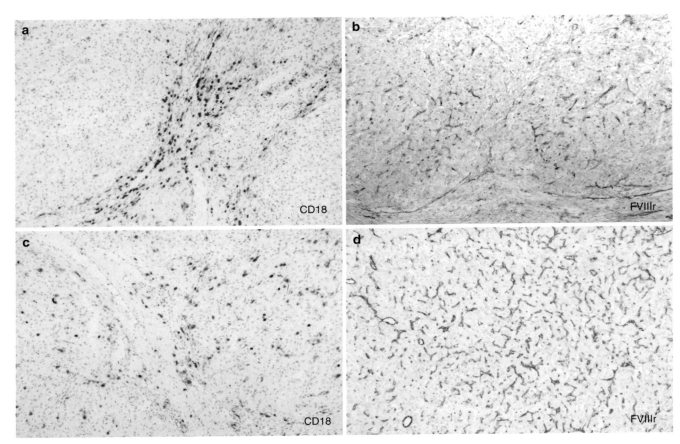

Fig. 6.7 Distribution patterns of leukocytes and changes of the microvascular bed in bovine CLs characterize different estrous cycle stages. Immunostaining for CD18 as general leukocyte marker is seen on the *left side*, and for FVIIr antigen to detect endothelial cells on the *right side*. (**a**, **c**, **e**, **g**) Leukocytes are dominant in the septum of a developing CL (**a**), uniformly distributed in the secretory stage (**c**), increasing in density at early regression, and clustered in the stage of advanced regression between prominent arterioles (**g**). (**b**, **d**, **f**, **h**) Capillaries are sprouting between the granulosa lutein cells in the stage of development (**b**). The capillary bed is well developed in the luteal tissue of secretion (**d**). Capillaries start disappearing and arterioles become prominent (**f**). Thick-walled arterioles represent a degenerative process (**h**). x a-h:70 (Adapted from Ricken et al. 1995; Spanel-Borowski et al. 1997)

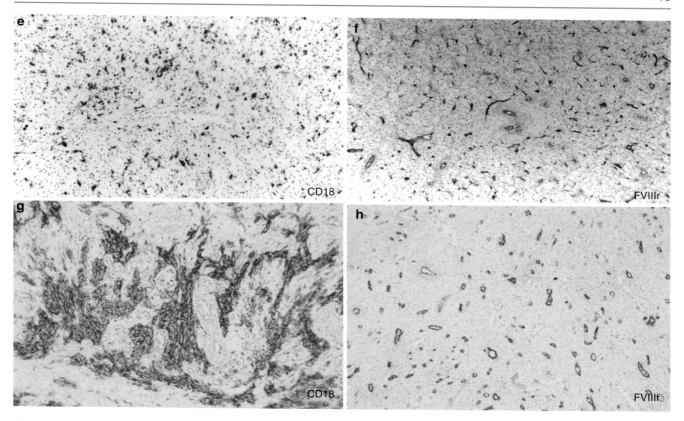

Fig. 6.7 (continued)

6.2.3   Eosinophils and Substance P-like Immunoresponse

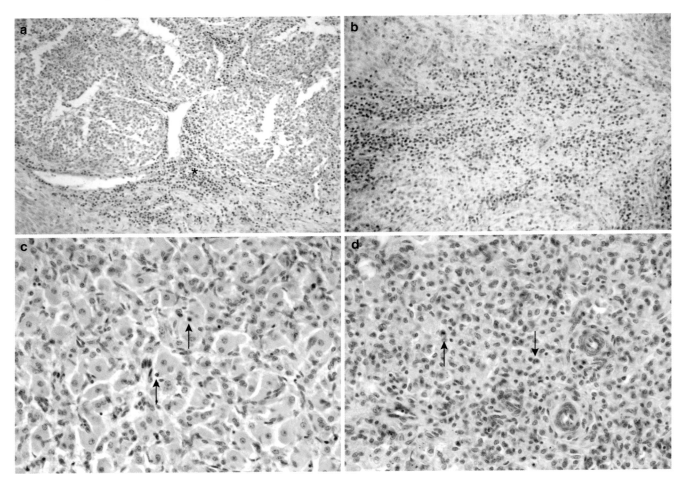

Fig. 6.8 Eosinophils are abundant in the bovine CL of early developmental stage as noted with Sirius red staining (**a–d**). The finding is associated with a high expression of a substance P-like immunoresponse (**e–h**). (**a–d**) In the stage of early development, the CL recruits many eosinophils in the former theca forming a septum (**a**, *asterisk*). Eosinophils accumulate close to a microvessel indicating their previous migration from the vessel lumen (**b**). In the secretory and regressing stages, a few eosinophils are distributed in the luteal tissue (*arrows* in **c**

and **d**). (**e**, **f**) In the early stage of development, arborizing forms with substance P-like immunoreactivity appear in the transition zone of the former theca and the luteinized granulosa. The immunoresponse might be of cellular origin (*arrow* in **f**). The *box* in **e** is enlarged in **f**. (**g**, **h**) Substance P-like fragments are uniformly distributed in the CL secretory stage (**g**). They are sparse and difficult to detect in the stage of regression (**h**). x a:30; b: 70; c,d.110; e:80; f,g,h:240 (Adapted from Reibiger and Spanel-Borowski 2000; Reibiger et al. 2001; Rohm et al. 2002)

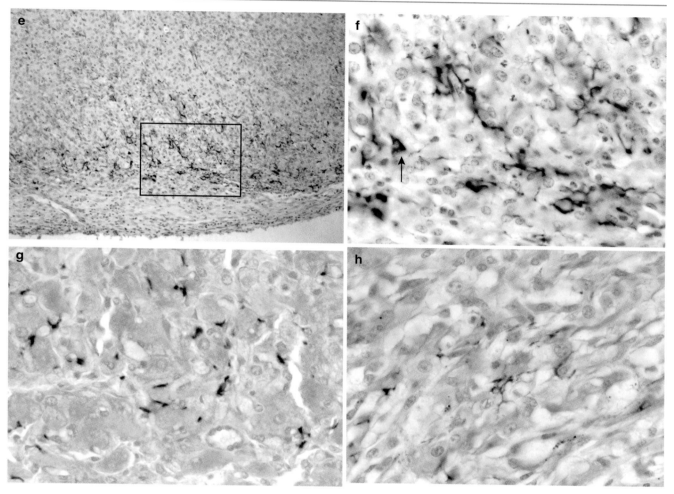

Fig. 6.8 (continued)

6.2.4   KIT-Positive Cells

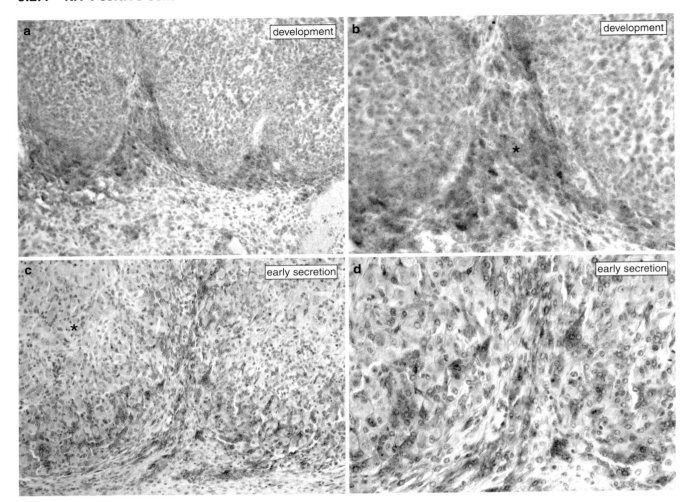

Fig. 6.9 KIT receptor-positive cells in the bovine CL are primarily theca-derived small luteal cells. Immunolocalization for positive cells is conducted with an anti-human polyclonal rabbit antiserum in section from different estrous cycle stages. (**a, b**) At the developmental stage, the prominent band-like staining in the very periphery of the CL coincides with the previous theca, which is forming the septum (*asterisk* in **b**). (**c, d**) At the early secretory stage, many KIT-positive cells are found in the septum, and they develop a network in the outer zone of the parenchyma, whereas the inner zone depicts a negligible response (*asterisk*). (**e–g**) At the late secretory stage, a few KIT-positive cells appear in the septum (*arrow*). The developed network consists of small KIT-positive luteal cells surrounding large negative cells (**e**). (**h**) At the stage of regression, a few KIT-positive cells with strong immunoreactivity look like leukocytes (*arrow*). The large luteal cells display a diffuse granular-like pattern. x a,c,e:120; b,d,f:240; g,h:380 (Adapted from Spanel-Borowski et al. 2007)

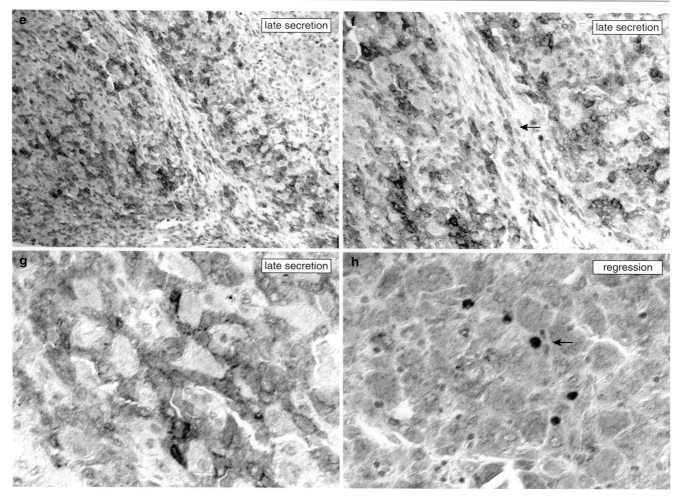

Fig. 6.9 (continued)

6.2.5   KIT-Positive Cells and CD45 Colocalization

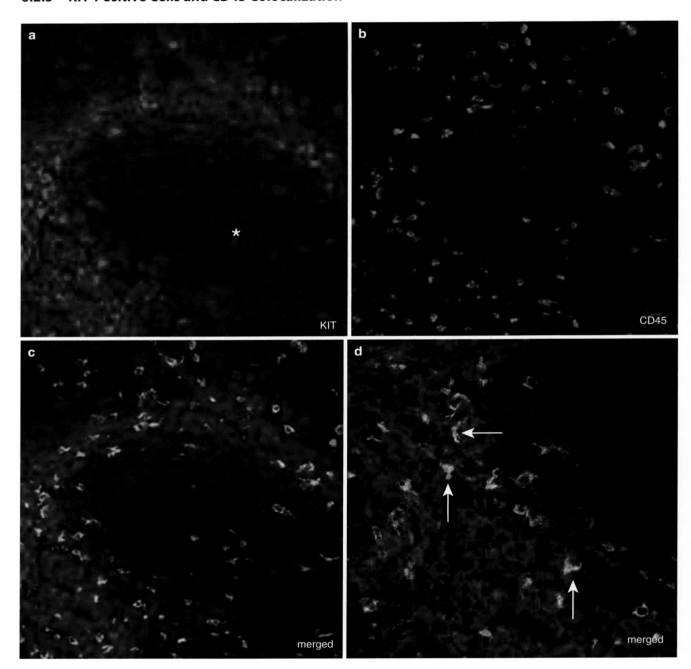

Fig. 6.10 The KIT-positive cell population in a bovine CL of early developmental stage comprises a minority of leukocytes positive for KIT and CD45 as detected in cryosections by double immunofluorescence localization of the previous theca. Staining was performed with a polyclonal anti-human antiserum against KIT and a monoclonal anti-bovine antibody against CD45, a leukocyte marker. KIT positive cells are in *red* (Cy3 fluorochrome for secondary antibody) and CD45 positive cells in *green* (Cy2 fluorochrome). (**a**) The band-like KIT positive response of the luteinized thecal layer is noted, whereas no signal appears in the luteinized granulosa (*asterisk*). (**b**) Leukocytes are distributed in the former theca. (**c**, **d**) Merged pictures depict orange cells as an overlap of KIT and CD45 (*arrows*). x a,b,c:180; c:360 (Adapted from Spanel-Borowski et al. 2007)

References

Bauer M, Reibiger I, Spanel-Borowski K (2001) Leucocyte proliferation in the bovine corpus luteum. Reproduction 121:297–305

Bauer M, Schilling N, Spanel-Borowski K (2003) Development and regression of non-capillary vessels in the bovine corpus luteum. Cell Tissue Res 311:199–205

Brylla E, Aust G, Geyer M, Uckermann O, Löffler S, Spanel-Borowski K (2005) Coexpression of preprotachykinin A and B transcripts in the bovine corpus luteum and evidence for functional neurokinin receptor activity in luteal endothelial cells and ovarian macrophages. Regul Pept 125:125–133

Choi J, Jo M, Lee E, Choi D (2011) The role of autophagy in corpus luteum regression in the rat. Biol Reprod 85:465–472

Debeljuk L (2006) Tachykinins and ovarian function in mammals. Peptides 27:736–742

Devoto L, Fuentes A, Kohen P, Cespedes P, Palomino A, Pommer R, Munoz A, Strauss JF 3rd (2009) The human corpus luteum: life cycle and function in natural cycles. Fertil Steril 92:1067–1079

Dunzendorfer S, Wiedermann CJ (2001) Neuropeptides and the immune system: focus on dendritic cells. Crit Rev Immunol 21:523–557

Gaytán F, Morales C, Bellido C, Aguilar R, Sanchez-Criado JE (2001) The fate of corpora lutea in the cyclic golden hamster. Gen Comp Endocrinol 121:104–113

Gaytán M, Morales C, Sanchez-Criado JE, Gaytán F (2008) Immunolocalization of beclin 1, a bcl-2-binding, autophagy-related protein, in the human ovary: possible relation to life span of corpus luteum. Cell Tissue Res 331:509–517

Gonçalves P, Ferreira R, Gasperin B, Oliveira J (2012) Role of angiotensin in ovarian follicular development and ovulation in mammals: a review of recent advances. Reproduction 143:11–20

Kaczmarek M, Schams D, Ziecik A (2005) Role of vascular endothelial growth factor in ovarian physiology - an overview. Reprod Biol 5:111–136

Klipper E, Levit A, Mastich Y, Berisha B, Schams D, Meidan R (2010) Induction of endothelin-2 expression by luteinizing hormone and hypoxia: possible role in bovine corpus luteum formation. Endocrinology 151:1914–1922

Medzhitov R (2010) Innate immunity: quo vadis? Nat Immunol 11:551–553

Meidan R, Levy N (2007) The ovarian endothelin network: an evolving story. Trends Endocrinol Metab 18:379–385

Munitz A, Levi-Schaffer F (2004) Eosinophils: 'new' roles for 'old' cells. Allergy 59:268–275

Nishigaki A, Okada H, Tsuzuki T, Cho H, Yasuda K, Kanzaki H (2011) Angiopoietin 1 and angiopoietin 2 in follicular fluid of women undergoing a long protocol. Fertil Steril 96:1378–1383

Nissim Ben Efraim AH, Levi-Schaffer F (2008) Tissue remodeling and angiogenesis in asthma: the role of the eosinophil. Ther Adv Respir Dis 2:163–171

Niswender G, Juengel JL, Silva PJ, Rollyson MK, McIntush EW (2000) Mechanisms controlling the function and life span of the corpus luteum. Physiol Rev 80:1–29

Reibiger I, Spanel-Borowski K (2000) Difference in localization of eosinophils and mast cells in the bovine ovary. J Reprod Fertil 118:243–249

Reibiger I, Aust G, Tscheudschilsuren G, Beyer R, Gebhardt C, Spanel-Borowski K (2001) The expression of substance P and its neurokinin-1 receptor mRNA in the bovine corpus luteum of early developmental stage. Neurosci Lett 299:49–52

Ricken AM, Spanel-Borowski K, Saxer M, Huber PR (1995) Cytokeratin expression in bovine corpora lutea. Histochem Cell Biol 103:345–354

Rohm F, Spanel-Borowski K, Eichler W, Aust G (2002) Correlation between expression of selectins and migration of eosinophils into the bovine ovary during the periovulatory period. Cell Tissue Res 309:313–322

Rönnstrand L (2004) Signal transduction via the stem cell factor receptor/c-Kit. Cell Mol Life Sci 61:2535–2548

Schams D, Berisha B (2004) Regulation of corpus luteum function in cattle–an overview. Reprod Domest Anim 39:241–251

Severini C, Improta G, Falconieri-Erspamer G, Salvadori S, Erspamer V (2002) The tachykinin peptide family. Pharmacol Rev 54:285–322

Spanel-Borowski K (2010) Weibliche genitalorgane. In: Zilles K, Tillmann BN (eds) Anatomie. Springer, Heidelberg, pp 546–569

Spanel-Borowski K (2011) Footmarks of innate immunity in the ovary and cytokeratin-positive cells as potential dendritic cells. Adv Anat Embryol Cell Biol, vol 209. Springer, Heidelberg. 978-3-642-16076-9

Spanel-Borowski K, Heiss C (1986) Luteolysis and thrombus formation in ovaries of immature superstimulated golden hamsters. Aust J Biol Sci 39:407–416

Spanel-Borowski K, Mayerhofer A (1987) Formation and regression of capillary sprouts in corpora lutea of immature superstimulated golden hamsters. Acta Anat (Basel) 128:227–235

Spanel-Borowski K, Rahner P, Ricken AM (1997) Immunolocalization of CD18-positive cells in the bovine ovary. J Reprod Fertil 111:197–205

Spanel-Borowski K, Amselgruber W, Sinowatz F (1987) Capillary sprouts in ovaries of immature superstimulated golden hamsters: a SEM study of microcorrosion casts. Anat Embryol 176:387–391

Spanel-Borowski K, Sass K, Löffler S, Brylla E, Sakurai M, Ricken AM (2007) KIT receptor-positive cells in the bovine corpus luteum are primarily theca-derived small luteal cells. Reproduction 134:625–634

Stocco C, Telleria C, Gibori G (2007) The molecular control of corpus luteum formation, function, and regression. Endocr Rev 28:117–149

Turvey SE, Broide DH (2010) Innate immunity. J Allergy Clin Immunol 125:S24–S32

Intraovarian Oocyte Release (IOR) with Severe Tissue Damage

The cyclic LH and FSH peak is well known to be responsible for the ovulatory event (Lunenfeld 2004; Tata 2005). Furthermore, the surface epithelium is said to support the degradation of the follicle wall (Wright et al. 2010). However, follicle rupture and intraovarian oocyte release (IOR) occur in preantral and antral follicles independent of gonadotropins and of the surface epithelium. IOR is a phenomenon seen in different species before and after puberty (Spanel-Borowski et al. 1982a, b, 1983a). Roughly ten IORs are found in completely cut ovaries of 19-day-old immature rats (Spanel-Borowski and Aumüller 1985) (Fig. 7.1c, d). As long as the oocyte is retained in the follicle antrum or at the rupture site, IOR is incomplete (Fig. 7.1a ,c ,e). After the oocyte is released into the cortical stroma, complete IOR has taken place (Fig. 7.1b, d, f). The IOR phenomenon often correlates with signs of atresia such as a deformed oocyte, ruptured zona pellucida, and granulosa cell subtypes like dark granulosa cells and albumin-positive cells (Fig. 7.2). The IOR process appears to be an inside-out event as has been suggested for the ovulatory process (Spanel-Borowski 2011a, b). Herniated granulosa cells show cytoplasmic shedding similar to interstitial gland cells or they look intact (Figs. 4.3 and 7.3).

The extracellular space of the IOR event is rich in defective mitochondria, rough endoplasmic membranes, lipid droplets, and vesicles. One wonders how the powerful organelles adapt to the new environment and how the extracellular space manages the invaders. The IOR oocyte contains annulate lamellae and cortical granules similar to a mature oocyte (Fig. 7.4a). Perhaps oocyte maturation has been accelerated in the IOR event. Microvessel lesions with discontinuities of the basement membrane could be a gateway to quickly remove remnants of the invaders (Fig. 7.4b, c). In the cortical stroma, the complete IOR from antral follicles causes typical after effects. A hyperchromatic cell group among less dense interstitial gland cells might contain an oocyte or a collapsed zona pellucida. Both structures are often not in continuity with the small IOR rupture site (Fig. 7.5a–d). The ruptured follicle can display hypertrophy of the thecal cell layer (Fig. 7.5a–f). In the case of IOR into a regressing CL,

one gets the impression that a nonruptured follicle has developed into a CL. Granulosa cells and a defective zona pellucida indicate a previous IOR into a CL (Fig. 7.6a–c). When a "naked" oocyte of healthy appearance and without apparent follicle cells is seen in the interstitial cortical stroma, a complete IOR into the cortical stroma could involve transition of follicle cells into interstitial gland cells (Figs. 7.6d and 3.2a–c).

The CL does not arise from preantral or antral IOR, but only from preovulatory IOR. The FSH and LH peak is needed to select dominant follicles out of the antral follicle cohort, which become preovulatory follicles. In superovulated ovaries, more preovulatory follicles develop than under physiological stimulation. Preovulatory follicles are crowded and obviously lose orientation toward the surface epithelium. Many preovulatory follicles undergo IOR with resumption of meiosis as shown for superovulated rabbits and rats (Figs. 7.7. and 7.8). In superovulated rabbit ovaries, roughly 20 preovulatory IORs per ovary are recorded 28 h after human chorionic gonadotropin (hCG) application. This indicates an extended time for the ovulatory event (Spanel-Borowski et al. 1986). It is noteworthy that preovulatory IOR is associated with CL appearance (Fig. 7.8c, d). Side effects like severe tissue damage are spectacular in preovulatory IOR in rats and rabbits (Spanel-Borowski et al. 1983b, 1986). Residues of the IOR follicle are found in the interstitial cortex often in a large edematous lake (Figs. 7.7c, d and 7.8e). Cumulus–oocyte complexes are released into damaged microvessels (Figs. 7.7e and 7.8g, h). Fibrin thrombi occur (Figs. 7.7e, f and 7.8h). The thrombi point to a change in the fibrinolytic activity before and after the ovulatory event (Spanel-Borowski and Heiss 1986). On day 4 after PMSG stimulation of rat ovaries, moderate fibrinolytic activity is limited to medullary microvessels and associated with a fresh fibrin thrombus, whereas on PMSG day 5, absence of activity relates to thrombi in organization (Fig. 7.9). The preovulatory IOR is obviously a dangerous event, which triggers the coagulating cascade of the complement system. This system represents the humoral protection side of INIM and acts

K. Spanel-Borowski, *Atlas of the Mammalian Ovary*,
DOI 10.1007/978-3-642-30535-1_7, © Springer-Verlag Berlin Heidelberg 2012

first by recognizing and transmitting threats into inflammatory responses (Köhl 2006; Matsushita 2009). Proteomic analysis of follicular fluid from women undergoing IVF points to the active involvement of the complement cascade in the ovulatory event (Jarkovska et al. 2010)

Pharmacological interference of IOR with melatonin treatment is presently of interest. Melatonin is known to mediate a pituitary-dependent antigonadotropic effect in hibernating animals. After treating adult golden hamsters with melatonin at 25 µg for 12 weeks, the majority of animals stopped cycling. This finding proves the antigonadotropic effect of melatonin (Spanel-Borowski et al. 1983a). Yet it remains difficult to understand why the melatonin-treated hamsters show 18 preantral and antral IOR per two ovaries/ one hamster in addition to extraovarian oocyte release (Fig. 7.10). The atypical follicle ruptures coincide with hypertrophy of thecal cells and of interstitial gland cells (Fig. 2.10e–g). Because LH blood levels are maintained in golden hamsters under melatonin treatment, cell hypertrophy could reflect an LH-dependent activation, thus an indirect effect causing atypical follicle ruptures. However, melatonin is recognized as a potent antioxidant that retards age-related increases in lipid peroxidation and acts directly on the immune system (Reiter et al. 1996; Okatani et al. 2002; Carrillo-Vico et al. 2006). In luteal cell cultures, melatonin stimulation overcomes the ROS-dependent inhibition of progesterone production, and, in women with luteal phase defect, melatonin treatment increases progesterone serum levels (Taketani et al. 2011). Thus, melatonin could exert a direct effect on follicles leading to IORs.

Arginine vasotocin (AVT), a nonapeptide occurring in the pineal gland, mediates antiovulatory function in the rat even though preovulatory peaks of LH, FSH, and prolactin are found (Spanel-Borowski et al. 1983b). The peptide thus acts at the level of the ovary. This is nicely depicted in superovulated ovaries of PMSG-treated immature rats (Fig. 7.11). While the PMSG-treated controls have about 20 IORs of antral and preovulatory type, four IORs are found in two ovaries/one rat of the PMSG-AVT group. The inhibition of IOR by AVT treatment (1 µg/0.1 ml every 2 h from 46 to 70 h after 30 IU PMSG stimulation at time "0") impairs neither the maturation of preovulatory follicles with cumulus expansion nor the resumption of meiosis. This indicates that gonadotropin and steroidogenic receptors are negligibly impaired by AVT treatment. In addition, the supporting role of prostaglandins in oocyte maturation appears to be unaffected (Spanel-Borowski et al. 1983b, 1986). Because AVT exerts a vasoconstrictive effect, the microvascular bed of the thecal cell layer seems to be the target of peptide treatment. Vasoconstriction is associated with decreased vessel permeability through which the flow of biomolecules is impeded in the thecal cell layer. Consequently, the thecal cell layer being a headquarter of capillary sprouting, of leukocyte recruitment, and of KIT up-regulation (Spanel-Borowski et al. 2007) is inhibited in its function of controlling follicle rupture. The inhibition might be delayed until AVT levels have become ineffective. A different type of rupture inhibition is described for repeated superovulations in golden hamsters (Löseke and Spanel-Borowski 1996). Transiently constricted thecal microvessels are not the cause, but rather a truly insufficient architecture of the microvascular bed (Fig. 5.5). The deficiency likely contributes to the persistence of dominant follicles (Fig. 7.12). No cumulus expansion and no resumption of meiosis is detected, in contrast to AVT treatment.

In summary, IOR occurs in different species such as small rodents, rabbits, and dogs. In spite of severe tissue damage by IOR side effects, superovulated ovaries return to full functionality at the onset of the next ovarian cycle. The mighty authority of INIM provides the capacity to degrade tissue by an acute inflammatory response and to coordinate tissue healing rapidly (Spanel-Borowski 2011a). These actions relate to the physiological side of INIM function. Insights into IOR morphology in superovulated ovaries are the basis for elucidating the molecular network of INIM signaling pathways, which control tissue damage and repair. Disturbances in the molecular cascade might be found in the ovary under melatonin application. Refractoriness to follicle rupture after AVT treatment or after repeated superovulations could also be associated with aberrations in INIM signaling.

7.1   IOR in Cyclic and Immature Ovaries

7.1.1   Incomplete and Complete IOR: Histology

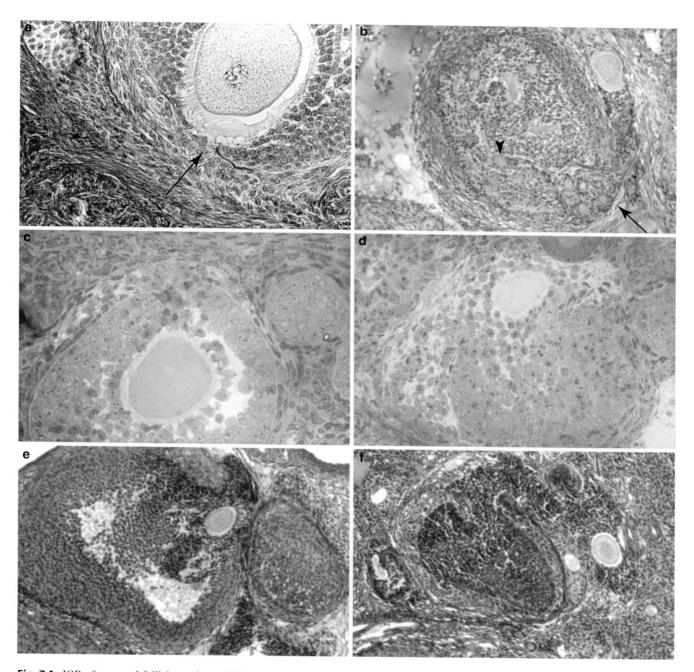

Fig. 7.1 IOR of preantral follicles and antral follicles is depicted for adult dogs (**a** and **b** after HOPA staining of paraffin sections), 19-day-old rats (**c** and **d** after Richardson staining of semithin sections), and mature rats (**e** and **f** after H&E staining). Incomplete IOR with the oocyte in the follicle compartment (*left side*). Complete IOR occurs with displacement of the oocyte outside the compartment (*right side*). (**a**) The preantral follicle with IOR reveals a partial discontinuity of the basement membrane close to the eccentric oocyte (*arrow*). (**b**) The antral follicle has released the deformed oocyte together with granulosa cells into the cortical stroma (*arrow*). The mural granulosa with Call–Exner bodies (*arrowhead*) is collapsed. (**c**) Incomplete preantral IOR is associated with slight deformation of the oocyte. (**d**) Complete preantral IOR displays herniated granulosa cells that look healthy. (**e, f**) Incomplete IOR and complete IOR occur in antral follicles. Herniated granulosa cells appear healthy. x a,c:300; b,e,f:100; d:220 (Adapted from Spanel-Borowski and Aumüller 1985; Spanel-Borowski et al. 1982a, 1983b)

7.1.2 Granulosa Cell Subtypes in IOR

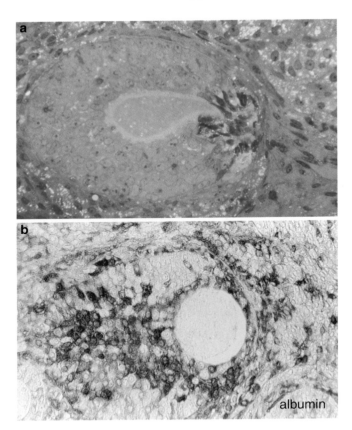

Fig. 7.2 The incomplete IOR of a preantral follicle is associated with the occurrence of dark cells and of albumin-positive cells. Semithin section with Richardson staining from the ovary of an immature 19-day-old rat (**a**) and paraffin section with immunohistology for albumin localization from an adult hamster (**b**). (**a**) The rupture area reveals dark cells within and outside of the follicle compartment. The oocyte is heavily deformed and a ruptured zona pellucida is seen as a sign of advanced atresia. (**b**) Albumin-positive granulosa cells are noted in incomplete preantral IOR. x a,b:300

7.1.3   Herniated Granulosa Cells: Ultrastructure

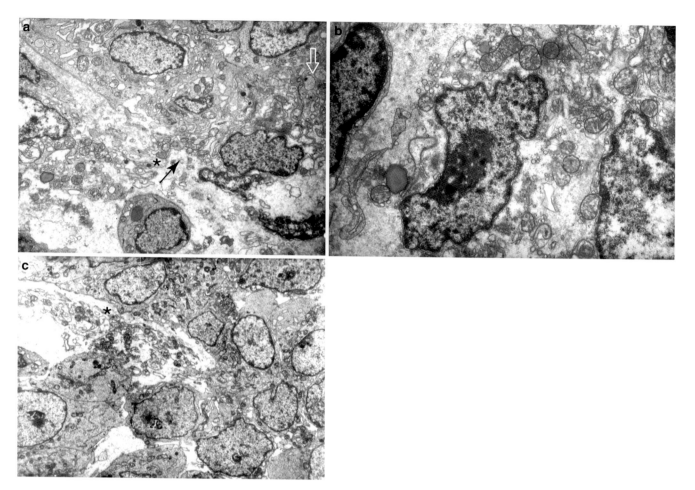

Fig. 7.3 The IOR rupture area is studied in ultrathin sections of 19-day-old rats. (**a, b**) Incomplete preantral IOR as in Fig. 7.1c reveals the partially destroyed basement membrane (*arrow in* **a**). Herniated granulosa cells disintegrate and large amounts of cytoplasmic contents appear in the extracellular space (*asterisk in* **a**). The herniated granulosa cell shows an irregular nucleus with a prominent nucleolus, while the cytoplasm is shed together with defective mitochondria (**b**). Mitochondria contain rarefied and damaged cristae with electron-dense intercristal space. Two annular junctions are barely discernible in intra-follicular granulosa cells (*open arrow* in **a**). (**c**) The complete preantral IOR as in Fig. 7.1d has expulsed intact granulosa cells with a vesicular nucleus through the ruptured basement membrane. Free organelles have accumulated in the extrafollicular space (*asterisk*). x a:3'600; b:8'600; c:2'800 (Adapted from Spanel-Borowski and Aumüller 1985)

7.1.4   IOR Oocyte and Microvessels: Ultrastructure

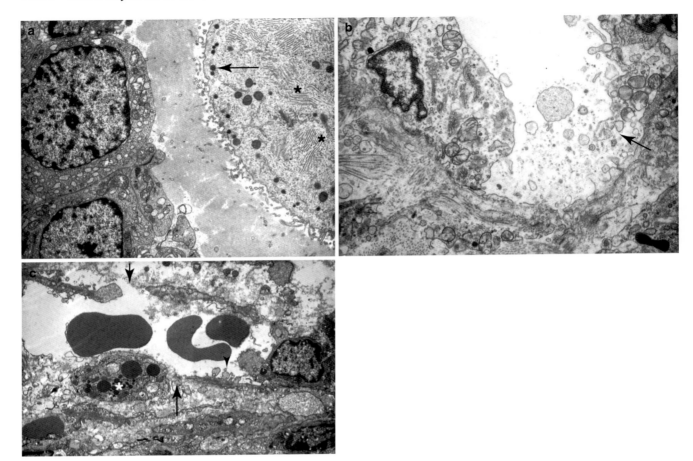

Fig. 7.4 The oocyte of an incomplete IOR as in Fig. 7.1c is studied in ultrathin sections of 19-day-old rats as are adjacent microvessels. (**a**) Numerous annulate lamellae (*asterisks*) and cortical granules (*arrow*) correspond to the structure of a mature oocyte. (**b**) The venule contains a high, activated endothelial cell. Another endothelial cell is disinte-grating (*arrow*). (**c**) The venule lacks the basement membrane in areas of endothelial fenestrations (*small arrow*) and in a wide gap (*large arrow*). It contains a cell fragment (*asterisk*). Endothelial cell blebbing is noted (*arrowhead*). x a:5'700; b:11'800; c:3'700 (Adapted from Spanel-Borowski and Aumüller 1985)

7.1.5   After Effects of IOR in the Cortical Stroma and in Corpora Lutea

7.1.5.1   Survival of Herniated Granulosa Cells

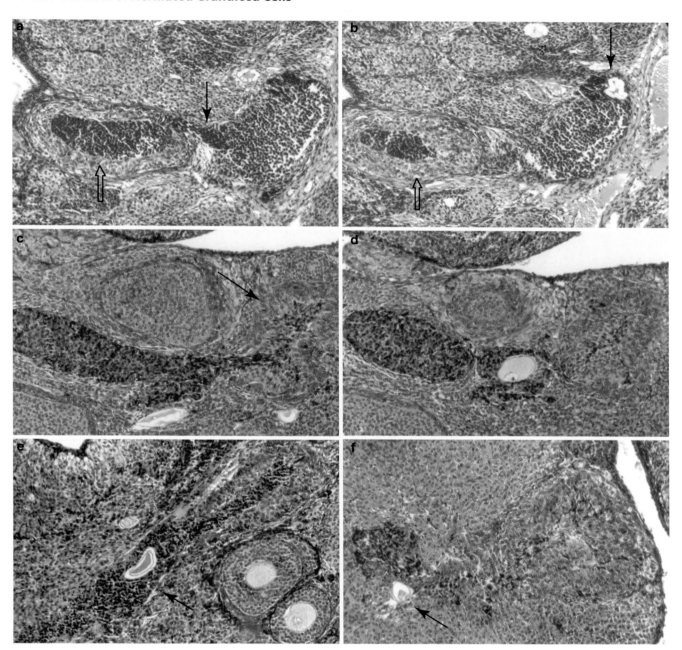

Fig. 7.5 After effects of IOR might escape detection without the study of serial sections here derived from mature rat ovaries (**a** and **b**) and golden hamsters treated with 25 μg melatonin for 12 weeks (**c–f**). Sections were stained with H&E. (**a, b**) The section plane reveals the IOR rupture site with herniated granulosa cells in (*arrow* in **a**). The subsequent section depicts the collapsed zona pellucida in the herniated cell group (*arrow* in **b**), which is far from the former follicle with hyper-trophy of the thecal layer (*open arrow*). (**c, d**) The herniated granulosa cells look like a tail originating from a collapsed follicle (*arrow* in **c**). The subsequent section shows the expulsed oocyte embedded in herni-ated granulosa cells. Contact to the follicle is lost (**d**). (**e, f**) Evidence of previous IORs is given in both section planes. Remnants of the oocyte and the zona pellucida are embedded in herniated granulosa cells (*arrow*), which are darker than the interstitial gland cells. x a-f:120

7.1.5.2 Residues of IOR in Corpora Lutea and in the Cortical Stroma

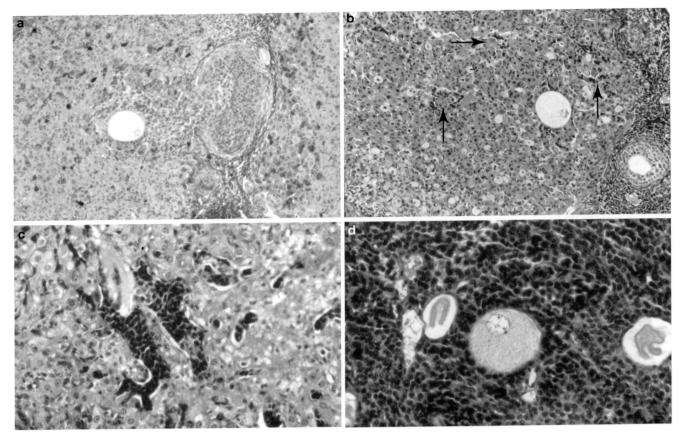

Fig. 7.6 After effects of IOR in the CL and in the stroma of the white-footed mouse (*Peromyscus leucopus*; **a**), of golden hamsters (**b, d**), and rats (**c**). Sections were stained with H&E. (**a**) Complete IOR with oocyte release into a regressing CL is noted. (**b**) The regressing CL encloses an oocyte and dissipated granulosa cells (*arrow*s). This indicates a previous IOR that escaped detection in the section plane. (**c**) A collapsed zona pellucida is associated with granulosa cells, which are disseminated in a CL. (**d**) A "free oocyte" resides in the interstitial cortical stroma. It looks healthy, and associated granulosa cells are barely discerned. x a,b:100; c,d:260 (Adapted from Spanel-Borowski et al. 1982a, 1982b)

7.2 Increase in IOR Through Superovulations and Side Effects

7.2.1 Severe Tissue Damage in Ovaries of Rabbits as Induced Ovulators

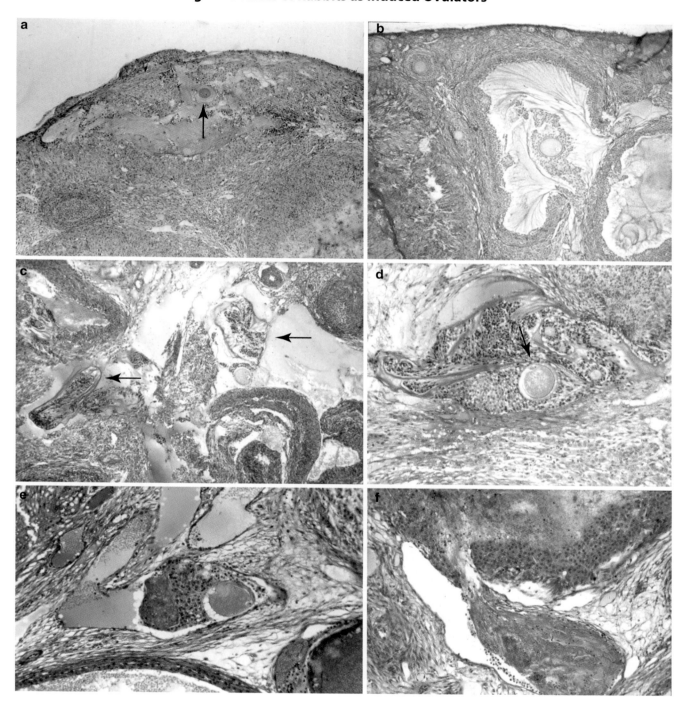

Fig. 7.7 Preovulatory follicles from superovulated rabbits undergo IOR that causes severe tissue damage as a side effect. Three-month-old rabbits were s.c. primed with 25 IU pure FSH per kg body weight at time "0," and ovulation was induced by an intravenous injection of 100 IU hCG/LH at 52 h. Rabbits were killed 15–48 h later. Serial sections were stained with HOPA. (**a**) The luteinized wall of a preovulatory follicle reveals four ruptures with leakage of follicular fluid into the cortical stroma. The oocyte is retained in the former antrum (*arrow*). (**b**) The incomplete IOR belongs to a large antral follicle without resumption of meiosis. (**c**) The IOR oocyte floats in edema containing remnants of the follicle wall (*arrows*). (**d**) Edema in the cortical stroma contains an IOR oocyte with the first polar body (*arrow*) indicating its origin from a preovulatory follicle with resumption of meiosis. (**e**) An oocyte–granulosa cell complex with a fibrin thrombus is seen in a dilated venule. (**f**) The IOR rupture is associated with a fibrin thrombus indicating increased fibrinolytic activity. x a,b,c:50; d,e,f:90 (Adapted from Spanel-Borowski et al. 1986)

7.2.2 Severe Tissue Damage in Rat Ovaries

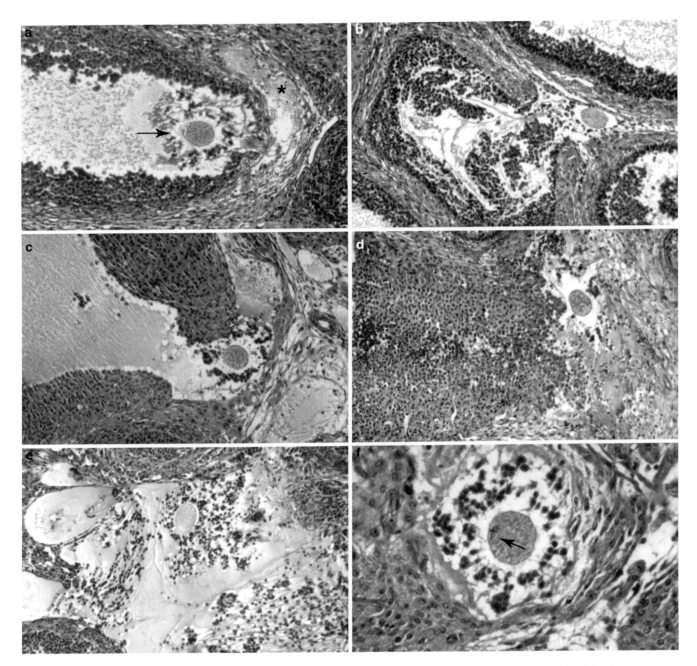

Fig. 7.8 Preovulatory follicles from superovulated rats undergo IOR that causes "false" CL formation and severe tissue damage as side effects. Twenty-seven-day-old rats were s.c. injected with 30 IU PMSG at time "0" and killed at 72 h. Serial sections were stained with H&E. (**a**) The incomplete IOR displays a polar body (*arrow*) and dispersed cumulus cells. Edema has developed at the rupture site (*asterisk*). (**b**) The complete IOR shows the oocyte–cumulus complex (COC) outside of the follicle compartment. Vesicle breakdown has occurred in the oocyte. (**c**) A CL cyst has originated from an incomplete IOR, as deduced from the COC at the rupture side. (**d**) A CL has developed after complete IOR. The expulsed oocyte lies in a hemorrhage. (**e**) A COC and dispersed granulosa cells float in a large edema of the interstitial cortical stroma. (**f**) A COC with a meiotic spindle (*arrow*) is located in the cortical stroma. (**g**) The COC with a deformed oocyte in a dilated venule speaks for an IOR rupture into a microvessel. (**h**) A fibrin clot with remnants of a COC is seen in a venule. x a,b,c,d,e,h:130; f,g:250 (Adapted from Spanel-Borowski et al. 1983b)

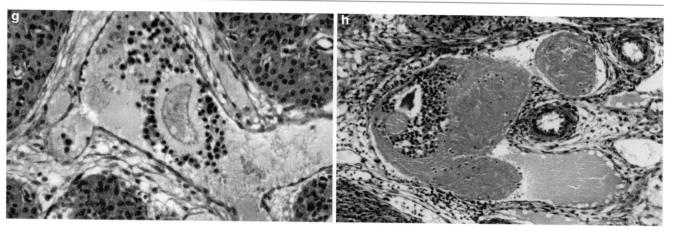

Fig. 7.8 (continued)

7.2.3   Thrombi in Golden Hamster Ovaries

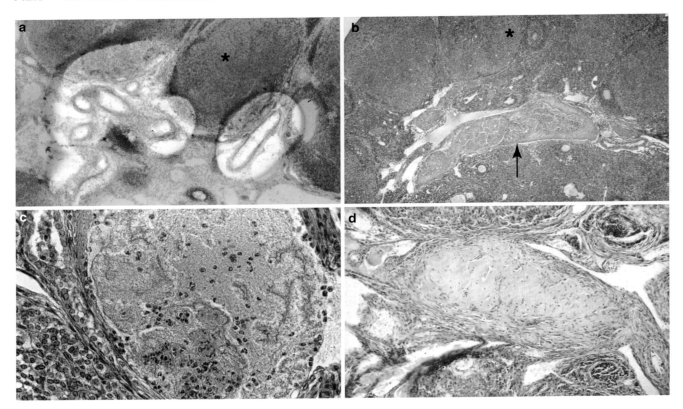

Fig. 7.9 Thrombus formation occurs in medullary vessels of supero-vulated golden hamsters. The process is associated with time-dependent changes of fibrinolytic activity. Twenty-eight-day-old hamsters were s.c. treated with 50 IU PMSG on day "0" followed by 25 IU hCG at 60 h. Serial sections were stained with H&E. (**a**) Moderate fibrinolytic activity occurs in medullary vessels on day 4 after PMSG application as detected using the fibrin slide method. No fibrinolytic activity is seen in the CLs (*asterisk* in **a**, compare with Fig. 5.1f). (**b, c**) A fresh fibrin thrombus has developed in a large medullary vessel on day 4 after PMSG application (*arrow* in **b**). Many CLs are crowded in the cortex (*asterisk* in **b**). (**d**) The thrombus undergoes organization on day 5 after PMSG application as deduced by ingrowth of connective tissue (*arrow*). x a,b:40; b:250; d:110 (Adapted from Spanel-Borowski and Heiss 1986)

7.3 Pharmacological Interference

7.3.1 IOR Increase by Melatonin

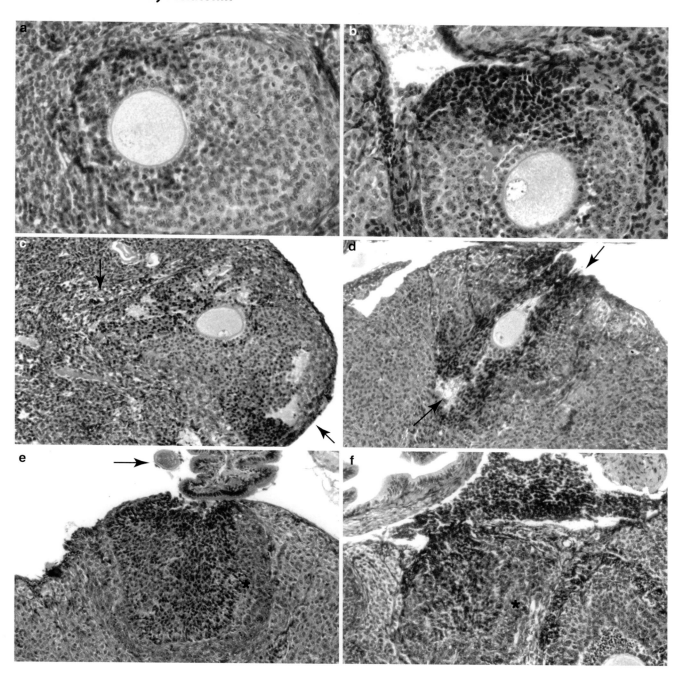

Fig.7.10 Melatonin (Mel) induces IOR and extraovarian oocyte release in golden hamsters. After daily injection with 15 μg Mel in 0.1 ml for 12 weeks, adult hamsters stopped cycling. Serial sections were stained with H&E. (**a**) Incomplete IOR involves a preantral follicle. (**b**) A preantral follicle has ruptured into a cleft of the surface epithelium documenting extraovarian oocyte release. (**c, d**) The antral follicles reveal two rupture sites going inside and outside of the ovary (*arrows*). The oocyte in the antrum is in diplotene arrest. The conspicu-ous hypertrophy of interstitial gland cells is evident (**d**). (**e, f**) The antral follicle has expulsed the oocyte (*arrow* in **e**) in the vicinity of the fallopian tube. Herniated granulosa cells lack luteinization in the subsequent section (**f**). The thecal cell layer shows distinct hypertrophy (*asterisk* in **e**). (**g**) The two collapsed antral follicles with former rupture site (*arrows*) indicate previous extraovarian oocyte release. (**h**) A "free" oocyte is seen. x a,b,h:250; c,d,e,f,g:130 (Adapted from Spanel-Borowski et al. 1983a)

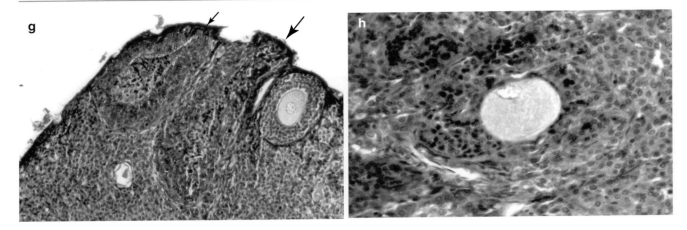

Fig. 7.10 (continued)

7.3.2   IOR Inhibition by Arginine-Vasotocin PMSG

PMSG

PMSG-AVT

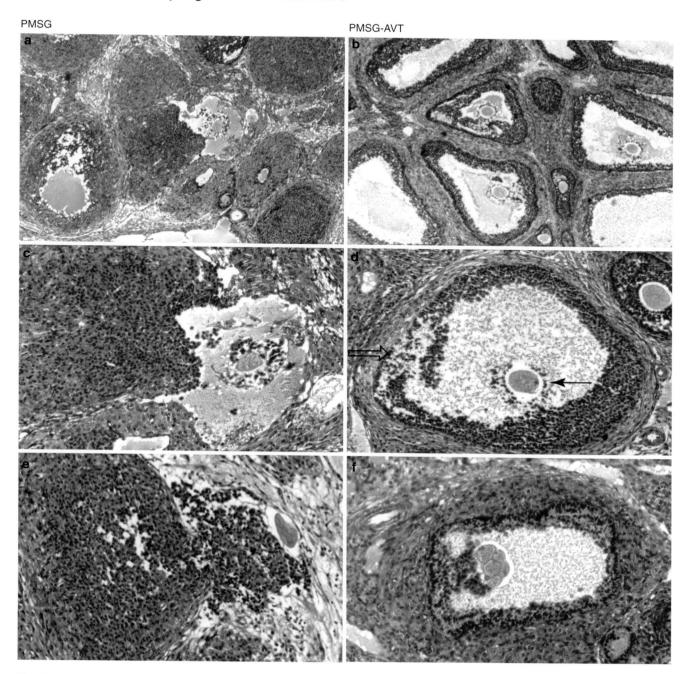

Fig. 7.11 Arginine-vasotocin (AVT) inhibits IOR in PMSG-primed immature rats, yet resumption of meiosis occurs. Twenty-seven-day-old rats were s.c. injected with 30 IU PMSG at time "0." Forty-six hours later 1 μg AVT/0.1 ml was s.c. injected every 2 h. The last injection was 70 h after PMSG application. Rats were killed 2 h later. Serial sections were stained with H&E. (**a, c, e**) Controls of PMSG-treated rats (*left column*) develop CLs partially attributed to IOR rupture. (**b, d, f**) In PMSG–AVT-treated rats (*right column*) preovulatory follicles persist. The mural granulosa layer is loosening (*open arrow* in **d**), the oocyte shows a meiotic figure (*arrow*) and a well-developed thecal cell layer. A morula-like form is noted (**f**). x a,b:60; c,d,e,f:120 (Adapted from Spanel-Borowski et al. 1983b)

7.3.3 IOR Inhibition by Repeated Superovulations

Once-Stimulated   Repeatedly stimulated

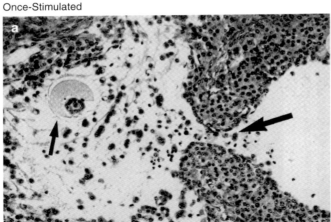

Fig.7.12 Repeated ovarian stimulations inhibit IOR from mature follicles. Thirty-day-old golden hamsters were stimulated by s.c. injections of 50 IU PMSG on day 0 and superovulations induced with 25 IU hCG at 60 h. Corpora lutea have disappeared on day 7 after PMSG application. Therefore, the subsequent seven superovulations were conducted on days 4, 14, 21 and so on after PMSG. Sections were stained with H&E. (**a**) In a once-stimulated ovary, complete IOR occurs (*small arrow*) and the rupture site (*large arrow*) of the luteinizing follicle is seen. (**b**) In the repeatedly stimulated hamster ovary, two large mature follicles are representative of IOR inhibition. Signs of meiosis are missing. x a:90; b:50 (Adapted from Löseke and Spanel-Borowski 1996)

References

Carrillo-Vico A, Reiter R, Lardone P, Herrera J, Fernandez-Montesinos R, Guerrero J, Pozo D (2006) The modulatory role of melatonin on immune responsiveness. Curr Opin Investig Drugs 7:423–431

Jarkovska K, Martinkova J, Liskova L, Halada P, Moos J, Rezabek K, Gadher SJ, Kovarova H (2010) Proteome mining of human follicular fluid reveals a crucial role of complement. J Proteome Res 9:1289–1301

Köhl J (2006) The role of complement in danger sensing and transmission. Immunol Res 34:157–176

Löseke A, Spanel-Borowski K (1996) Simple or repeated induction of superovulation: a study on ovulation rates and microvessel corrosion casts in ovaries of golden hamsters. Ann Anat 178:5–14

Lunenfeld B (2004) Historical perspectives in gonadotrophin therapy. Hum Reprod Update 10:453–467

Matsushita M (2009) Ficolins: complement-activating lectins involved in innate immunity. J Innate Immun 2:24–32

Okatani Y, Wakatsuki A, Reiter R, Miyahara Y (2002) Melatonin reduces oxidative damage of neural lipids and proteins in senescence-accelerated mouse. Neurobiol Aging 23:639–644

Reiter RJ, Pablos MI, Agapito TT, Guerrero JM (1996) Melatonin in the context of the free radical theory of aging. Ann N Y Acad Sci 786:362–378

Spanel-Borowski K (2011a) Footmarks of innate immunity in the ovary and cytokeratin-positive cells as potential dendritic cells. Adv Anat Embryol Cell Biol, vol. 209. Springer, Heidelberg ISBN 978-3-642-16076-9

Spanel-Borowski K (2011b) Ovulation as danger signaling event of innate immunity. Mol Cell Endocrinol 333:1–7

Spanel-Borowski K, Aumüller G (1985) Light and ultrastructure of intra-ovarian oocyte release in infantile rats. Anat Embryol (Berl) 172:331–337

Spanel-Borowski K, Heiss C (1986) Luteolysis and thrombus formation in ovaries of immature superstimulated golden hamsters. Aust J Biol Sci 39:407–416

Spanel-Borowski K, Petterborg LJ, Reiter RJ (1982a) Preantral intra-ovarian oocyte release in the white-footed mouse, peromyscus leucopus. Cell Tissue Res 226:461–464

Spanel-Borowski K, Vaughan MK, Johnson LY, Reiter RJ (1982b) Occurrence of preantral intra-ovarian oocyte release in the rat and the Syrian hamster (*Mesocricetus auratus*). Anat Embryol (Berl) 165:169–175

Spanel-Borowski K, Richardson BA, King TS, Petterborg LJ, Reiter RJ (1983a) Follicular growth and intraovarian and extraovarian oocyte release after daily injections of melatonin and 6-chloro-melatonin in the Syrian hamster. Am J Anat 167:371–380

Spanel-Borowski K, Vaughan LY, Johnson LY, Reiter RJ (1983b) Increase of intra-ovarian oocyte release in PMSG-primed immature rats and its inhibition by arginine vasotocin. Biomed Res 4:71–82

Spanel-Borowski K, Sohn G, Schlegel W (1986) Effects of locally applied enzyme inhibitors of the arachidonic acid cascade on follicle growth and intra-ovarian oocyte release in hyperstimulated rabbits. Arch Histol Jpn 49:565–577

Spanel-Borowski K, Sass K, Löffler S, Brylla E, Sakurai M, Ricken AM (2007) KIT receptor-positive cells in the bovine corpus luteum are primarily theca-derived small luteal cells. Reproduction 134:625–634

Taketani T, Tamura H, Takasaki A, Lee L, Kizuka F, Tamura I, Taniguchi K, Maekawa R, Asada H, Shimamura K, Reiter R, Sugino N (2011) Protective role of melatonin in progesterone production by human luteal cells. J Pineal Res 51:207–213

Tata J (2005) One hundred years of hormones. EMBO Rep 6:490–496

Wright JW, Pejovic T, Lawson M, Jurevic L, Hobbs T, Stouffer RL (2010) Ovulation in the absence of the ovarian surface epithelium in the primate. Biol Reprod 82:599–605

Vascular Stromatolysis and Vascular Luteolysis for Bulk Tissue Removal

Cell proliferation, cell maintenance, and cell degradation regulate tissue homeostasis. The complex molecule interactions in control of cell proliferation are still an important research topic for understanding organ development, malignancies and their therapy (Nordman and Orr-Weaver 2012; Canavese et al. 2012). Success in therapy can be attributed to an increase in apoptosis, which has been studied for more than two decades. Nonapoptotic cell death forms are presently attracting the attention of researchers as novel avenues of tissue homeostasis (Tsujimoto 2012). The possibility of physiological tissue disintegration by bulk tissue removal has been unknown . It is here described for immature golden hamster ovaries strongly stimulated with a protocol of 50 IU PMSG and 25 IU LH. The ovarian weight decreases from 160 to 30 mg between PMSG days 4 and 7, the former being associated with the presence of many CLs and the latter with their complete absence (Spanel-Borowski and Heiss 1986). There are two explanations for the striking decrease in weight of the superovulated hamster ovaries. Firstly, abundant apoptosis causes acute luteolysis and complete removal of CLs (Figs. 6.1 and 6.2). Secondly, the process of vascular stromatolysis emerges as a novel idea to explain the quick reduction in the amount of cortical tissue (unpublished results). In fact, complexes of mature interstitial gland cells are seen to protrude into cortical microvessels obviously through damaged vessel walls (Figs. 8.1a–d and 8.2). They likely give way to albumin diffusion into the extravascular space and thus explain the diffuse albumin-positivity located by immunostaining in the vicinity of cortical vessels (Fig. 8.1e, f). Vascular stromatolysis can be considered an intelligent mechanism for rapidly reducing the amount of cortical tissue by instantaneous removal of interstitial gland cells. As described in Chap. 4, interstitial gland cells show full steroidogenic activity during the ovulatory period (Mossman and Duke 1973; Guraya 1978). The well-developed endocrine cells provide androgen as a substrate for increased estrogen synthesis to the preovulatory follicle. After ovulation, a minor androgen input seems to be immediately regulated by vascular stromatolysis in superovulated hamster ovaries. Vascular stromatolysis is judged as a novel mechanism of rapid tissue disappearance with the aim of establishing cellular balance at once. Vascular stromatolysis is distinguished from cellular stromatolysis by cytoplasmic shedding in canine interstitial gland cells described as a novel form of cell degeneration (Chap. 4 and Fig. 4.3). Thus, homeostasis of interstitial gland cells varies in regulation with respect to cell number and cell size.

Vascular stromatolysis in golden hamster ovaries compares with vascular luteolysis in white-footed mice (Spanel-Borowski et al. 1983). The white-footed mouse with a 4-day-cycle maintains inactive CLs from previous estrous cycles (Margolis and Lynch 1981). Histological snapshots depict CL hemorrhage, disappearance of endothelial cells, and mobilization of luteal cell complexes into the supporting microvessel (Fig. 8.3). The complexes are subsequently noted as thrombi in medullary vessels (Fig. 8.4). Vascular luteolysis occurs in mice kept under short and long photoperiods to a similar extent, although more persistent CLs are found in mice kept under a short photoperiod than under a long one. Vascular luteolysis can be increased in intensity by intravenous prolactin injections (30 μg twice daily for 2–6 days). Vascular luteolysis in white-footed mice differs from structural luteolysis. At the onset, dark luteal cells provide long cell processes between light luteal cells (Fig. 8.5a, b). The dark cell could represent a dendritic-like cell (Chap. 10). When a few apoptotic bodies appear, endothelial cells are no longer apparent (Fig. 8.5c, d). In the final stage of structural luteolysis, luteal cells are crowded with lipid droplets and contain hemosiderin granules (Fig. 8.5e, f). They testify to previous hemorrhages into the CL, recalling vessel wall damage during vascular luteolysis. Hence, vascular and structural luteolysis are coexistent in white-footed mice (Fig. 8.6) and most likely occur in other mammalian CLs as well. In support of this statement are old findings with hypophysectomized rats describing the disappearance of persistent CLs within days after prolactin injections (Malven and Sawyer 1966; Malven et al. 1969). The classic concept of structural luteolysis states that apoptotic steroidogenic and vascular luteal cells

K. Spanel-Borowski, *Atlas of the Mammalian Ovary*,
DOI 10.1007/978-3-642-30535-1_8, © Springer-Verlag Berlin Heidelberg 2012

are phagocytosed through macrophages and replaced by ingrowth of connective tissue (van Wu et al. 2004). This concept has some shortcomings. The redundant demand to master balanced removal of the CL from the previous ovarian cycles can hardly be executed by heterophagocytosis alone. Autophagy is as an additional option for handling apoptotic luteal cells (Choi et al. 2011). Furthermore, structural luteolysis can be classified into an immediate process with complete disappearance of CLs in golden hamsters within 2 days (Figs. 6.1 and 6.2). The extended process of structural luteolysis requires several ovarian cycles as is known for rats and cows (Fig. 6.7). It is conceivable that both ways of structural luteolysis hide subunits of vascular stromatolysis.

The question arises of why vascular stromatolysis and vascular luteolysis have escaped our notice until today. The answer is that these unique processes surpass our present understanding of tissue degradation. Simply the expected and the known are realized, according to an old saying. Scientists generally study projects based on a solid scientific background and avoid taking risks with "obscure" theories. Someone in the near future might have the intuition to encounter here an innovative observation with far-reaching implications. Appropriate molecular techniques are needed to meet the scientific challenge of deciphering the molecular steps of vascular luteolysis. The need for molecular evidence of vascular luteolysis carries with it an important vision, because the CL life cycle compares with tumor development and its spontaneous healing. The immune system, in particular INIM, can be the driving force both in vascular luteolysis and in spontaneous tumor healing.

8.1   Vascular Stromatolysis in Histology

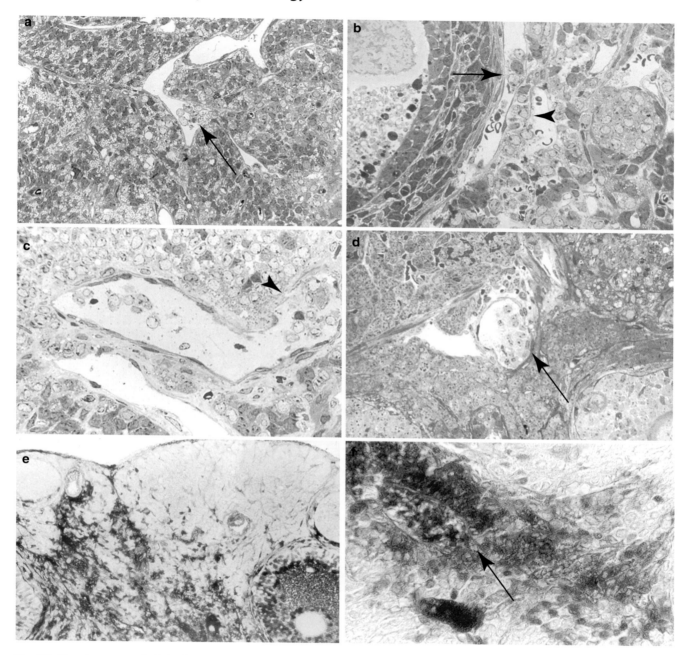

Fig. 8.1 Vascular stromatolysis is observed in ovaries of 28-day-old golden hamsters on day 6 after PMSG application. Hamsters were s.c. injected with 50 IU PMSG at time "0" and 25 IU hCG at 60 h. (**a–d**) The stroma consists of mature and resting interstitial gland cells in Richardson-stained semithin sections (Chap. 4). Microvessels contain cells that are reminiscent of interstitial gland cells (*arrows*). The continuity of the vessel wall can be disrupted (**b** and **c**, *arrow* *head*). (**e, f**) Immunolocalization of albumin reveals areas that are both poor and rich in protein. This finding might reflect leakage of albumin through a ruptured vessel wall (*arrow* in **f**). The antral follicle in the lower right corner shows albumin positivity in granulosa cells (Chap. 10) and in the follicular fluid as positive control in e. x a,d:190; b,c:300; e:110; f:220

8.2   Vascular Stromatolysis in Ultrastructure

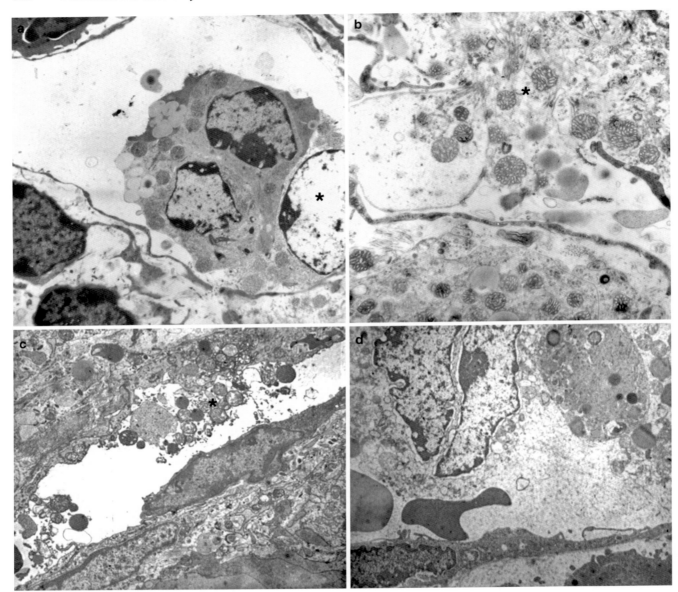

Fig. 8.2 Vascular stromatolysis is detected in ultrathin sections derived from superovulated ovaries of 28-day-old golden hamsters on day 6 after PMSG application. (**a**) The microvessel of the cortical stroma contains steroidogenic cells with lipid droplets and tubular-type mitochondria. A nucleus undergoes lysis (*asterisk*). (**b**, **c**) The moment of interstitial gland cell segregation through the ruptured wall of a microvessel (*asterisk*) is captured. (**d**) Interstitial gland cells disintegrate in the vessel lumen under the aspect of cytoplasmic shedding (see Chap. 4). x a:4'200; b:6'700; c:4'000; d: 4'300

8.3   Vascular Luteolysis with Prolapse of Luteal Tissue

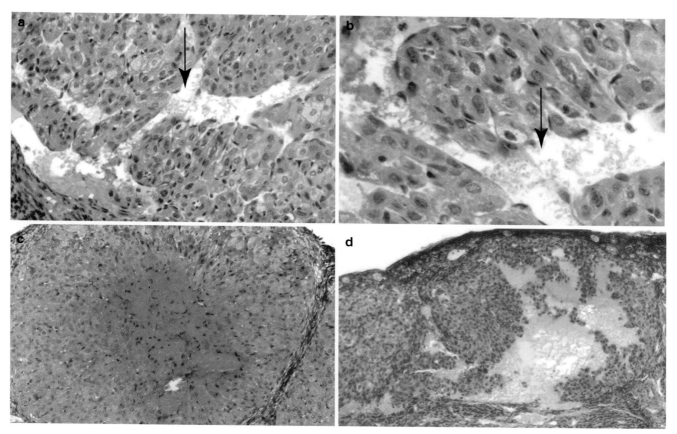

Fig. 8.3 Vascular luteolysis is described as an acute event of luteolysis seen in H&E-stained sections of adult white-footed mouse (*Peromyscus leucopus*) ovaries. (**a**, **b**) The supporting microvessel of a CL is dilated and partially lacks endothelial cell lining. Luteal cells without evident degenerative changes come into contact with the vessel lumen (*arrow*). (**c**) A hemorrhage disrupts the integrity of a CL. (**d**) Fragmentation of the CL could be caused by a severe hemorrhage. x a: 240; b:480; c,d:120 (Adapted from Spanel-Borowski et al. 1983)

8.4 Vascular Luteolysis with Luteal Cell Thrombi

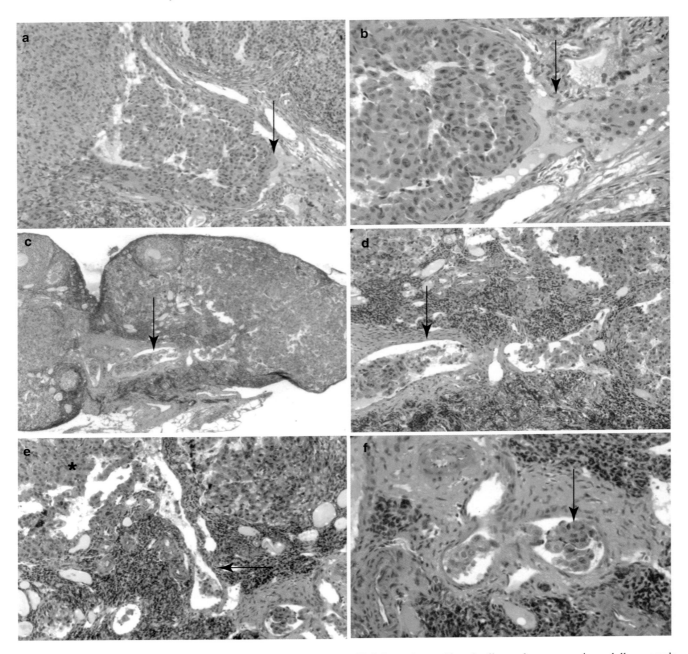

Fig. 8.4 Vascular luteolysis in the adult white-footed mouse (*Peromyscus leucopus*) is associated with luteal cell thrombi in medullary vessels of the ovary as shown after H&E staining. (**a, b**) A fragment of a regressing CL is segregated into a venule (*arrow*). (**c, d**) The CL is loosening and luteal cell complexes appear in medullary vessels (*arrow*). (**e, f**) Fragmentation of the CL (*asterisk* in **e**) is associated with luteal cell thrombi in longitudinal/cross-cuts of microvessels (*arrow*). x a,d,e:120; b,f:240; c:50 (Adapted from Spanel-Borowski et al. 1983)

8.5   Extended Structural Luteolysis with Fatty Degeneration of Luteal Cells

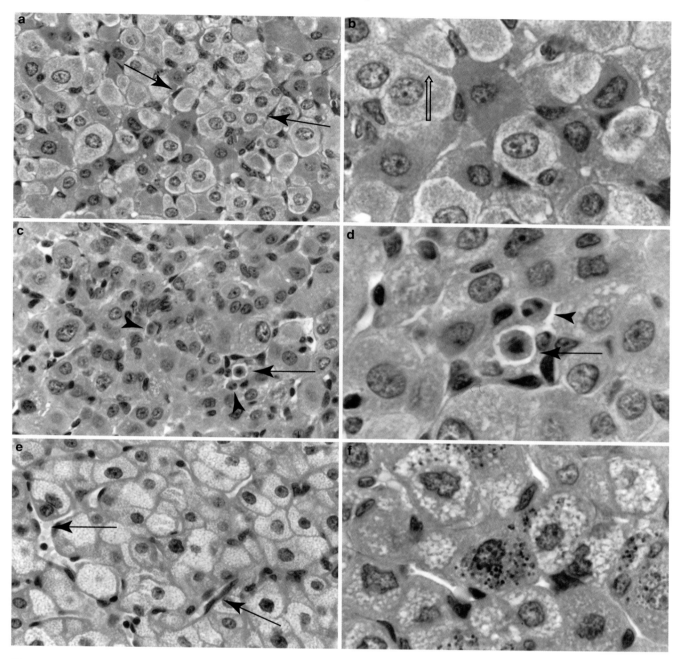

Fig. 8.5 Structural luteolysis with apoptosis, eosinophil infiltration, and fatty degeneration occurs in parallel to vascular luteolysis (Figs. 8.3 and 8.4) as shown in H&E-stained sections of the adult white-footed mouse (*Peromyscus leucopus*) ovary. (**a, b**) The change of the CL from the secretory into the regressing stage is still associated with the presence of endothelial cells *(arrows* in **a**) and of two luteal cell types. The light luteal cell possesses a faintly stained eosinophilic cytoplasm. The dark cell with striking eosinophilia extends long cell processes between light cells *(open arrow* in **b**). This finding supports the concept that dendritic-like luteal cells have a CL function as described in Chap. 10. (**c, d**) Apoptotic bodies *(arrow)* together with bilobed leukocytes/eosinophils *(arrow heads)* indicate early structural luteolysis. The light and dark luteal cell types have disappeared. (**e, f**) Fatty degeneration of luteal cells represents the final stage of structural luteolysis. Of note, capillaries are maintained *(arrows* in **e**). Brown granules in (**f**) could be lipofuscin or hemosiderin. x a,c,e:480; b,d,f:1050 (Adapted from Spanel-Borowski et al. 1983)

8.6   Persistent Corpora Lutea with Hemosiderin Deposits

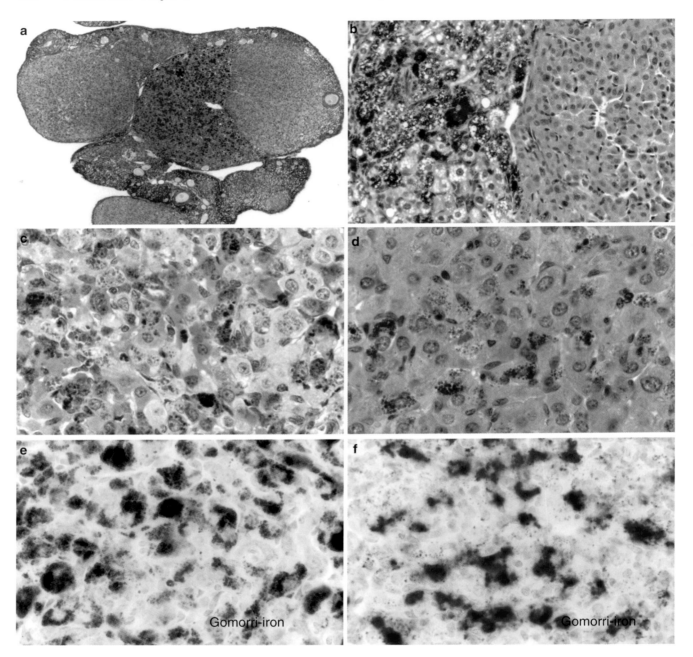

Fig. 8.6 Adult white-footed mice (*Peromyscus leucopus*) kept under short photoperiods for 6 weeks maintain CLs in different stages of regression. Sections were stained with H&E (**a–d**) and with Gomorri-iron solution (**e, f**). (**a, b**) The dark-brown CL (*asterisk*) is rich in lipofuscin aggregates, which accumulate in luteal cells under fatty degeneration (**b**). The adjacent CL consists of eosinophilic luteal cells without the lipochrome. (**c**) Light and dark luteal cells have accumu- lated aggregates of lipofuscin. (**d**) The brown granules in some eosino- philic luteal cells might indicate iron deposits after hemorrhage during vascular luteolysis. (**e, f**) Validation of brown granules and aggregates in CLs gives a negative/brown response (lipofuscin) and a positive/blue one (hemosiderin), respectively. The striking reaction product in *blue* (**f**) speaks for a heavy hemorrhage during vascular luteolysis. a:50; b,e,f:240; c,d.480

References

Canavese M, Santo L, Raje N (2012) Cyclin dependent kinases in cancer: Potential for therapeutic intervention. Cancer Biol Ther 13: 451–457

Choi J, Jo M, Lee E, Choi D (2011) The role of autophagy in corpus luteum regression in the rat. Biol Reprod 85:465–472

Guraya SS (1978) Recent advances in the morphology, histochemistry, biochemistry, and physiology of interstitial gland cells of mammalian ovary. Int Rev Cytol 55:171–245

Malven PV, Sawyer CH (1966) A luteolytic action of prolactin in hypophysectomized rats. Endocrinology 79:268–274

Malven PV, Cousar GJ, Row EH (1969) Structural luteolysis in hypophysectomized rats. Am J Physiol 216:421–424

Margolis DJ, Lynch GR (1981) Effects of daily melatonin injections on female reproduction in the white-footed mouse, peromyscus leucopus. Gen Comp Endocrinol 44:530–537

Mossman H, Duke K (1973) Comparative morphology of the mammalian ovary. Univ. Madison Wisconsin Press, Madison

Nordman J, Orr-Weaver T (2012) Regulation of DNA replication during development. Development 139:455–464

Spanel-Borowski K, Heiss C (1986) Luteolysis and thrombus formation in ovaries of immature superstimulated golden hamsters. Aust J Biol Sci 39:407–416

Spanel-Borowski K, Bartke A, Petterborg LJ, Reiter RJ (1983) A possible mechanism of rapid luteolysis in white-footed mice, peromyscus leucopus. J Morphol 176:225–233

Tsujimoto Y (2012) Multiple ways to die: Non-apoptotic forms of cell death. Acta Oncol 51:293–300

van Wu R, der Hoek KH, Ryan NK, Norman RJ, Robker RL (2004) Macrophage contributions to ovarian function. Hum Reprod Update 10:119–133

Polycystic and Postmenopausal Ovaries with Negligible Mast Cells

Mast cells belong to the cellular defense system of INIM (Turvey and Broide 2010; Medzhitov 2010a, b). Mast cell presence in the ovary could thus point to areas with INIM functions in respect to innate or acquired immunoresponses (Galli et al. 2011; Shelburne and Abraham 2011). In the bovine ovary, toluidine blue-stained mast cells appear to be absent in the sparsely vascularized cortical rim, in follicles, and in CLs (Figs. 4.1c, 5.11f, and 6.6d). Their absence could signify the minor contribution of mast cells to the life of follicles and of CLs. Mast cells are uniformly distributed in the interstitial cortical stroma and in the medulla where they represent roughly half of the CD18-positive cell pool. The percentage is consistent throughout different ovarian cycle stages in the bovine ovary (Reibiger and Spanel-Borowski 2000). Mast cells might be influential in the turnover of nerve fibers in the cortex and the medulla, because they produce neurotrophic factors that could mediate a balanced neuroimmune connection in the cyclic ovary. An impaired neuroimmune interaction is demonstrated in the dehydroepiandrosterone (DHEA)-induced polycystic syndrome in rat ovaries (Krishna et al. 2001). The DHEA-treated ovaries lack toluidine blue mast cells in the medulla, while the density of nerve fibers positive for calcitonin gene-related peptide (CGRP) is increased (Fig. 9.1). The disappearance of mast cells could signify a subacute cell degranulation rendering mast cells undetectable with immunostaining. The release of neurotrophic factors through degranulation might partially explain the increase in nerve fiber density. The DHEA-induced cystic follicles seem to be another source for the production of neurotrophic factors such as neurite growth factor (NGF) (Dissen et al. 2009).

The distribution of mast cells is investigated in the cyclic human ovary. Tryptase-positive mast cells avoid the cortical rim and the antral follicle wall. They prefer to populate the hilus, the medulla, and the corticomedullary region (Fig. 9.2a–c). It is noteworthy that the significant decrease and disappearance of mast cells in ovaries with polycystic syndrome and of postmenopausal women do not correlate with a decrease in leukocytes (Fig. 9.2e–h). It is therefore suggested that either maturation of young mast cells is impeded or degranulation becomes subacute (Heider et al. 2001; Krishna et al. 2001) Membrane–membrane contacts between mast cells and peptidergic/S100-positive nerve fibers are rarely seen in the cortex of the cyclic human ovary (Fig. 9.3). The reason is that areas with many mast cells develop few nerve fibers, and the reverse situation also occurs (Fig. 9.3a). The membrane–membrane contacts allude to a bi-directional communication such as the release of pro-NGF from nerve fibers and the mast cell-dependent cleavage of pro-NGF to NGF through tryptase (Ito and Oonuma 2006; Spinnler et al. 2011). Mast cells become inapparent in immunostained sections of polycystic ovaries and of postmenopausal ovaries, while S100-positive nerve fibers increase in density (Fig. 9.4). The fibers depict discontinuities and varicosities (Fig. 9.4c, d). Of note, the scarcity of nerve fibers in the cortex periphery of the cyclic ovary is reminiscent of the sparsely vascularized cortical rim and the mast-cell-free zone in the bovine ovary (Figs. 4.1a, b, 9.2a, and 9.4b). Under the aspect of INIM control in the ovary, mast cell presence correlates with the maintenance of the ovarian cycle, thus with the cyclic maturation of interstitial gland cells. As suggested in Chap. 4, the role of mast cells is limited to supervising cellular stromatolysis in such a way that shedding of cytoplasmic components into the extracellular space avoids any inflammatory reaction and thus occurs as a nonresponsive event. When the life cycle of interstitial gland cells ceases in postmenopausal ovaries, the protective function of mast cells is cancelled and no further mast cells are recruited.

9.1   Polycystic Rat Ovaries After Dehydroepiandrosterone (DHEA) Treatment

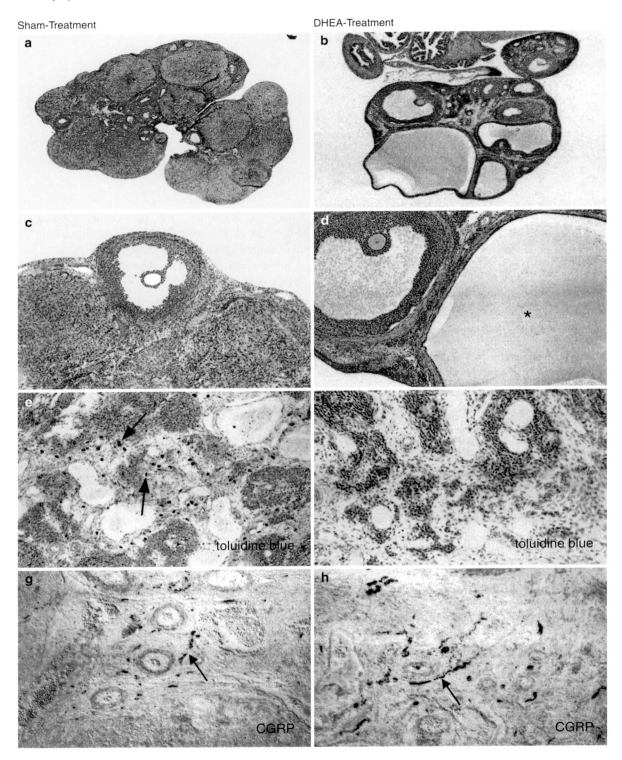

Fig. 9.1 Peptidergic nerves are increased and mast cells decreased in DHEA-induced polycystic rat ovaries. Thirty-one-day-old rats were s.c. injected with 6 mg DHEA per day for 30 days. Sham-treated ovaries are on the *left* side and compared with DHEA-treated ovaries on the *right* side. Sections (**a–d**) were stained with H&E, (**e, f**) and with toluidine blue. (**g, h**) Immunohistology for calcitonin gene-related peptide (CGRP) was performed. (**a, b**) The control ovary is larger because of many CLs, which are not seen in the DHEA-treated ovaries. They contain large antral follicles often of cystic appearance. (**c, d**) The antral follicle of a control ovary is smaller than that of a DHEA-treated one. The follicle cyst (*asterisk* in **d**) that occurs under DHEA treatment indicates pathway B of follicular atresia (Chap. 2). (**e, f**) Mast cells populate the ovarian medulla of untreated/control ovaries. Mast cells are absent in DHEA-treated ovaries. (**g, h**) In controls, short CGRP-positive fibers are revealed between blood vessels of the medulla. In the DHEA-treated ovaries, the CGRP-positive fibers of the medulla appear to be thicker and longer (*arrow*). x a:30; b:80; c,d,e,f:60; g:100; h:120 (Adapted from Krishna et al. 2001)

9.2 Mast Cells and Leukocytes in Human Ovaries

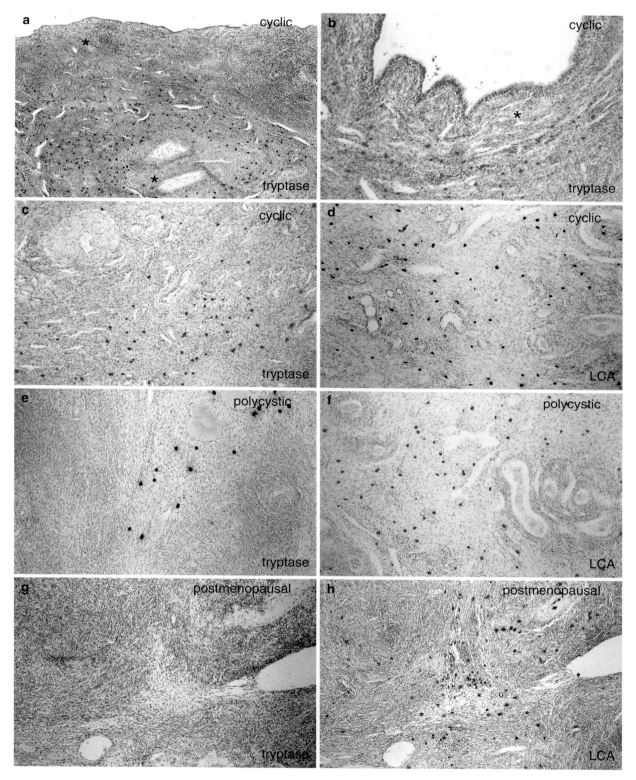

Fig. 9.2 Mast cells and leukocytes are studied in human ovaries in the follicular phase (**a–d**), compared with polycystic ovaries (**e, f**) and postmenopausal ovaries with hyperthecosis (**g**, **h**), respectively. Immunostaining was conducted for tryptase (**a–c, e, g**) and for common leukocyte antigen (LCA) (**d, f, h**). (**a, b**) In the cyclic ovary, mast cells are absent in the outer cortex and in regressing follicles (*asterisks*). (**c,** **d**) In the corticomedullary region, the number of mast cells appears to be similar to the number of LCA-positive leukocytes. (**e–h**) In polycystic ovaries (**e, f**) and in postmenopausal ovaries (**g, h**), tryptase-positive mast cells decrease or become absent in the corticomedullary region. Leukocyte density compares with the cyclic ovary. x a:35; b,c,d,f,g:80; e:120 (Adapted from Heider et al. 2001)

9.3   Membrane–Membrane Contacts Between Mast Cells and S100-Positive Nerve Fibers

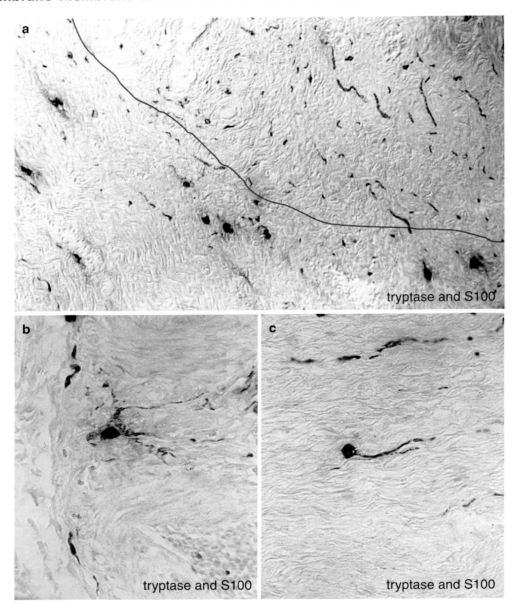

Fig. 9.3 Mast cells and peptidergic nerve fibers are co-localized in human ovaries in the follicular phase by double immunolabeling for tryptase-positive mast cells (*brown*) and for S100-positive nerve fibers (*red*). (**a**) The area with preferential nerve fibers is separated from an area with mast cells (dashed line for separation). Membrane–membrane contacts between mast cell and nerve fiber are limited to the transition zone between both areas. (**b, c**) A neuromast cell contact shows mast cells in *brown* and nerve fiber in *red*. A neuron-like structure is seen in (**b**). x a:110; b,c:160 (Adapted from Heider et al. 2001)

9.4   **Different Densities of S100-Positive Nerve Fibers**

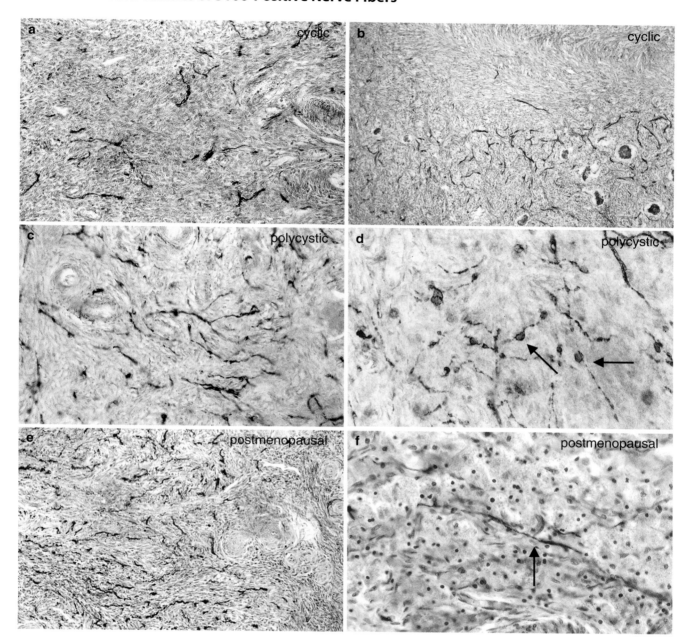

Fig. 9.4 Peptidergic nerve fibers are inferior in density in human ovaries in the follicular phase (**a**, **b**) compared with polycystic ovaries (**c**, **d**) and postmenopausal ovaries (**e**, **f**). Single immunolabeling for S100 antigen was performed. (**a**) S100-positive nerve fibers are displayed for the corticomedullary region. (**b**) The outer cortex of cyclic ovaries is poor in S100-positive nerve fibers similar to the scarcity of mast cells (compare Fig. 9.2a). (**c**, **d**) More S100-positive nerve fibers appear in the polycystic ovary than the cyclic ovary in (**a**). Nerve fibers show discontinuities in the immunoresponse and varicosities (*arrows* in **d**). (**e**, **f**) The highest nerve fiber density occurs in the postmenopausal ovary. Fibers look thick and touch interstitial gland cells (*arrow*) in an area with hyperthecosis. x a,c,e,f:230; b:80; d:240 (Adapted from Heider et al. 2001)

References

Dissen G, Garcia-Rudaz C, Paredes A, Mayer C, Mayerhofer A, Ojeda S (2009) Excessive ovarian production of nerve growth factor facilitates development of cystic ovarian morphology in mice and is a feature of polycystic ovarian syndrome in humans. Endocrinology 150:2906–2914

Galli S, Borregaard N, Wynn T (2011) Phenotypic and functional plasticity of cells of innate immunity: macrophages, mast cells and neutrophils. Nat Immunol 12:1035–1044

Heider U, Pedal I, Spanel-Borowski K (2001) Increase in nerve fibers and loss of mast cells in polycystic and postmenopausal ovaries. Fertil Steril 75:1141–1147

Ito A, Oonuma J (2006) Direct interaction between nerves and mast cells mediated by the SgIGSF/SynCAM adhesion molecule. J Pharmacol Sci 102:1–5

Krishna A, al Rifai A, Hubner B, Rother P, Spanel-Borowski K (2001) Increase in calcitonin gene related peptide (CGRP) and decrease in mast cells in dihydroepiandrosterone (DHEA)-induced polycystic rat ovaries. Anat Embryol 203:375–382

Medzhitov R (2010a) Inflammation 2010: new adventures of an old flame. Cell 140:771–776

Medzhitov R (2010b) Innate immunity: quo vadis? Nat Immunol 11:551–553

Reibiger I, Spanel-Borowski K (2000) Difference in localization of eosinophils and mast cells in the bovine ovary. J Reprod Fertil 118:243–249

Shelburne C, Abraham S (2011) The mast cell in innate and adaptive immunity. Adv Exp Med Biol 716:162–185

Spinnler K, Frohlich T, Arnold G, Kunz L, Mayerhofer A (2011) Human tryptase cleaves pro-nerve growth factor (pro-NGF): hints of local, mast cell-dependent regulation of NGF/pro-NGF action. J Biol Chem 286:31707–31713

Turvey SE, Broide DH (2010) Innate immunity. J Allergy Clin Immunol 125:S24–S32

Cytokeratin-Positive Cells as Dendritic-Like Cells and Potential Albumin Coexpression in Follicles and CL

Follicle cells represent a heterogeneous cell population in morphology and function. A few examples are as follows. In the preovulatory follicle, the oocyte-associated cell layer stops cell proliferation and undergoes columnar metaplasia and cumulus expansion (Figs. 2.2c, d and 5.9d). The difference in response is explained by different membrane receptor expression comparing the inner and the outer/mural cell layer. In follicle harvests from IVF patients, granulosa cells with angiogenic potential are likely related to endothelial precursor cells (Antczak and Van Blerkom 2000; Merkwitz et al. 2010). The harvests also allow isolation of CK-positive follicle cells as a unique subpopulation. They first appear in the genital ridge during the early gestational period and are derived from the CK-positive coelomic epithelium and the CK-positive sex cord epithelial cells, both densely populated with CK-negative germ cells (Fig. 10.1a–d). The medullary part of the sex cords segregates primordial follicles with CK-positive follicle cells (Fig. 10.1e, f). From childhood to puberty, CK-positive follicle/granulosa cells transiently disappear during the growth from primary to secondary follicles (Fig. 10.2a–c). CK-positive granulosa cells preferentially reappear in the basal layers of antral follicles and are intensified in number in the preovulatory follicle (Fig. 10.2d–f). Hence, the characteristic pattern of CK expression differs in the fetal and the postfetal period, as is schematically summarized in Fig. 10.3 (Löffler et al. 2000). The maximum of CK expression occurs in the ruptured follicle giving way to "zonation" in the CL of early developmental stage. Zonation is caused by the infolded granulosa cell layer, which becomes prominent as a band-like structure of granulosa lutein cells with strong CK expression (Fig. 10.3a, b). The CK-positive zones are separated by infoldings of the thecal cell layer, which consists of small thecal lutein cells without CK expression (Ricken et al. 1995). In the stage of secretion, the pattern of zonation is lost in the CL, because large lutein cells with strong, moderate, and negligible CK expression are uniformly distributed. Additionally, small CK-positive cells with strong CK positivity are noted. By double-staining for CK and for laminin as basement membrane markers, small

CK-positive cells are located in the microvascular bed. Thus, after follicle rupture, CK expression is at its maximum in large lutein cells and subsequently decreases to a minimum between the stages of secretion and regression (Fig. 10.4c–f). Small CK-positive vessel-associated cells are an exception. The bovine CL during pregnancy lacks any CK-positive steroidogenic cells. As is concluded, the functional state of the CL decides on the presence or absence of steroidogenic CK-positive cells in the bovine and human CL (Ricken et al. 1995). This conclusion finally settles the extensive debate at the end of the twentieth century on the presence and significance of CK-positive cells in the CL (Czernobilsky et al. 1985; Santini et al. 1993). Because abundant CK-positive follicle cells are found in the pre-and postovulatory follicle and because immunostaining intensities for CK decrease in the CL of secretory stage, a switch-on and switch-off action is assumed for CK genes. This possibility is shown for transformed cells, which, in vitro, lose the complex partner CK8 together with a rapid degradation of the CK18 protein (Knapp and Franke 1989). The switching on of CK genes is likely independent of LH, but caused by intraovarian factors, which are activated by oxidative stress as a final motor of the ovulatory event (Spanel-Borowski 2011a, b). The statement is substantiated by the appearance of CK-positive granulosa cells in regressing antral follicles, which are known to lack LH receptors, but undergo inflammatory-like changes (Figs. 3.3 and 3.5f–h). They are considered to be a response to oxidative stress in analogy to similar changes in preovulatory follicles. Further support for the switch-on and switch-off concept comes from an overlooked finding that the expression of CK filaments in granulosa cells depends on culture condition (Ben-Ze'ev and Amsterdam 1989). Switching off of CK genes and thus of CK-filament-dependent intercellular contacts apparently provides mobility to granulosa cells, which is needed at the time of transformation from a preovulatory follicle into a CL.

The switch-on and switch-off concept of CK expression in granulosa/luteal cells is important for the present classification of five different phenotypes of endothelial cell

K. Spanel-Borowski, *Atlas of the Mammalian Ovary*,
DOI 10.1007/978-3-642-30535-1_10, © Springer-Verlag Berlin Heidelberg 2012

cultures derived from the bovine CL (Spanel-Borowski and van der Bosch 1990; Spanel-Borowski 1991). The five phenotypes compare with different monolayers of epithelial cells (Fig. 10.5) (Fenyves et al. 1993). Type 1 develops a prominent cobble-stone pattern of uniform cells, which contain a nice CK-positive cytoskeleton. The monolayer of type 2 shows irregular cell forms, caused by the co-culture of CK-positive endothelial cells with desmin-positive/CK-negative vascular cells. Types 3 and 4, either spindle-shaped or polygonal, are classic microvascular endothelial cells with many endothelial cell-specific FVIIIr-positive granules. Type 5 represent the largest cells forming a flat monolayer with difficult-to-discern nuclei. The weak presence of 3β HSD activity and the change into a multilayer in the presence of endothelial cell growth factors finally define type 5 cells as granulosa-like cells (Spanel-Borowski et al. 1994a, b). It is suggested that type 5 cells originate from CK-positive steroidogenic luteal cells that have lost CK filaments in cultures without LH (Spanel-Borowski 2011a, c). Type 1, however, represents the CK-positive microvascular cell type that maintains CK expression also in the CL during the stages of secretion and of regression. The endothelial cell classification of type 1 relies on extensive studies at the mRNA and protein level, which have revealed endothelial cell-specific criteria in the absence of steroidogenic production (Aust et al. 1999; Tscheudschilsuren et al. 2002). Type 1 lacks receptors for FSH and LH at the mRNA level. In contrast to endothelial cells that die under IFN-γ treatment (200 U/0.5 ml for 3 days), types 1 and 5 survive and display characteristics similar to dendritic cells under cytokine treatment. Treated cells of types 1 and 5 proliferate (Fenyves et al. 1994). Both types develop a peripheral band of actin filaments in contact with cellular junctions. Tight junctions and adherens junctions increase under IFN-γ treatment (Ricken et al. 1996). The CD18 adhesion molecule is augmented, as reflected by increased leukocyte adhesion to the monolayer surface (Ley et al. 1992). The single cilium characterizes type 1 as a danger sensor more than type 5, which develops a ciliary stub (Wolf and Spanel-Borowski 1992). The nonmotile single cilium is presently judged as a sensory organelle with signaling capacity (Davis et al. 2006; Singla and Reiter 2006). Type 5 extends very long filipodia toward neighboring cells and up-regulates MHC class II 80-fold compared to the sevenfold up-regulation of type 1 (Spanel-Borowski and van der Bosch 1990; Spanel-Borowski and Bein 1993). Provided that types 1 and 5 compare with dendritic-like cells, they belong to the cellular arm of INIM function in the CL (Spanel-Borowski 2011a, c). The postulated switch-on and switch-off in CK expression as assumed for steroidogenic luteal cells explains the characteristic CK patterns in the different CL stages and the origin of types 1 and 5 (Fig. 10.6). Although final proof is needed, dendritic-like cells in the ovary with the capacity to alter CK expression are part of an innovative and revolutionary concept of ovulation and luteolysis (Spanel-Borowski 2011a, b).

Because endothelial cell types 1 and 5 in the CL have likely originated from follicles, these types might be related to granulosa cell subtypes in preantral and antral follicles. Therefore, the appearance of albumin-positive cells in follicles and in CLs of mice is examined more closely. Albumin carries a variety of substances, for example, steroids and lipids, and delivers them from the serum to activated tissues where the ligand-receptor complex recognizes specific membrane-binding sites. Thereafter, the complex is internalized by endocytosis and metabolized by the formation of phagolysosomes. Pharmacologists use the principle to deliver therapeutic agents to sites of inflammation and tumor (Neumann et al. 2010; Elsadek and Kratz 2012). In the ovary, albumin-positive cells are localized in follicles and in CLs of rats and golden hamsters (Fig. 10.7), which could indicate the internalization of an albumin-receptor complex for the delivery of steroids and lipids to activated cells (Spanel-Borowski 1987; Krishna and Spanel-Borowski 1989). Distinct albumin-positivity in oocyte-associated granulosa cells of rat preantral follicles is judged as the onset of follicular atresia, because it compares with the sudden appearance of CK-positive cells in regressing bovine follicles (Figs. 10.7a and 3.5e, f). It is speculated that albumin-positive cells represent CK-positive cells appearing under atresia-related oxidative stress. The same might hold true for preovulatory rat follicles depicting strong albumin-positivity in the cumulus oophorus and the inner granulosa cells. Albumin-positive cells in the basal granulosa cell layer seem to partially coincide with CK expression in the basal area of bovine antral follicles (Figs. 10.2c and 10.7c–f). Likewise the CL of golden hamsters shows vascular cells either with a strong albumin-immunoresponse or with a weak one. Two subtypes behave similar to CK-negative and CK-positive vascular cells in the bovine CL (Figs. 10.4e, f and 10.7g, h). Our hypothesis that the expression of albumin and CK in ovarian cells identifies comparable cells is further substantiated by the complete IOR associated with herniation of albumin-positive granulosa cells into the cortical stroma (Fig. 10.8). The herniated cells look intact, thus excluding passive uptake of albumin by dying cells. The increase in albumin-positive granulosa cells in combination with IOR basement membrane rupture resembles the increase in CK-positive cells in the ruptured bovine follicle (Figs. 10.4a and 10.8).

CK-positive granulosa cell cultures from human follicles up-regulate PRRs such as TLR4 under oxLDL treatment, as recently described (Serke et al. 2009, 2010). It is known that TLR-related signaling requires co-regulatory receptors such as CD14 for TLR4 activation (Miller et al. 2003). In macrophages, CD36 is the co-receptor to recognize oxLDL before the assembly of TLR4 and 6 into a heterodimer (Stewart et al. 2010). The membrane co-receptor is influential proximal to the cascade activation by the TLR heterodimer complex. The cascade itself can be modified by intracellular interactions with the complement pathway (Köhl 2006). The presence of varying co-receptors and interactions appears to fine-tune the complex TLR inflammatory pathways as either TRIF-dependent or Myd88-dependent (Takeuchi and Akira 2010; O'Neill and Bowie 2007). The albumin membrane-binding site might recognize the oxLDL–albumin complex prior to the potential interaction with TLR4 leading to internalization of the complex. These ideas are highly speculative. The working hypothesis should be validated by co-localization of CK and albumin in granulosa cell subtypes and in vascular cells of the CL using ovarian sections from the same species. Provided co-localization is successful, albumin and its binding sites are suggested as a general marker for selecting immunocompetent cells in ovaries of different species.

10.1   CK-Positive Cells in the Fetal Ovary

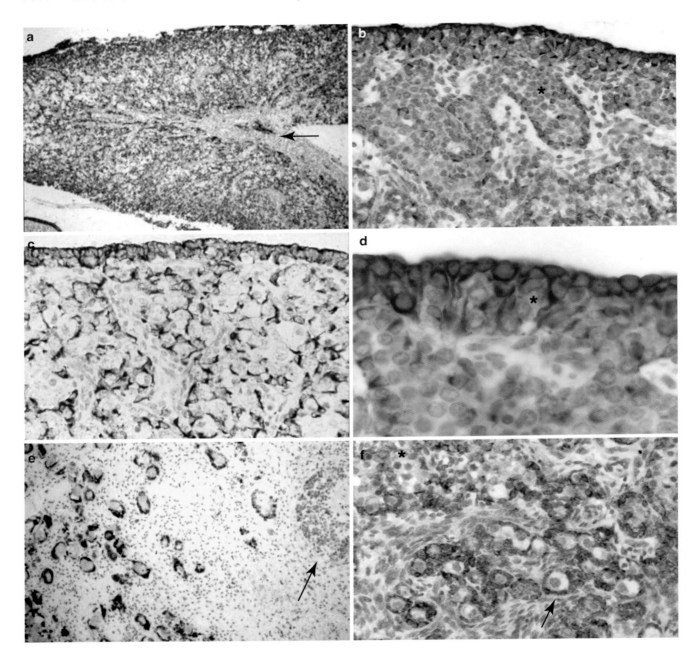

Fig. 10.1 CK-positive cells are present in ovaries from a human fetus (**a**, **c**, **e**, *left side*) and from a bovine fetus (**b**, **d**, **f**, *right side*). Immunohistology was done with the pan-CK antibody Lu5. (**a**, **c**) The human ovary in the early gestational trimester shows CK-positivity in the surface/coelomic epithelium, in sex cord epithelial cells, and in rete tubules of the medulla (*arrow* in **a**). (**b**, **d**) In the bovine ovary from a fetus of 28 cm crown–rump length, the delicate CK-positive network represents the CK-positive epithelial cells of sex cords. They enclose germ cells in first meiotic stages also seen in the surface/coelomic epithelium (*asterisks*). (**e**) In the late gestational period, CK-positive follicle cells (*left*) belong to primordial follicles, not to the preantral follicle (*arrow*). (**f**) In analogy, sections of the bovine ovary (**b**, **d**) show the medullary part of sex cords (*asterisk*), which segregate primordial follicles with the oocyte in diplotene arrest (*arrow*). x a.60; b,c,f:240; d:420; e:130 (Adapted from Löffler et al. 2000; Tsikolia et al. 2009)

10.2   Transient Disappearance During Follicular Growth

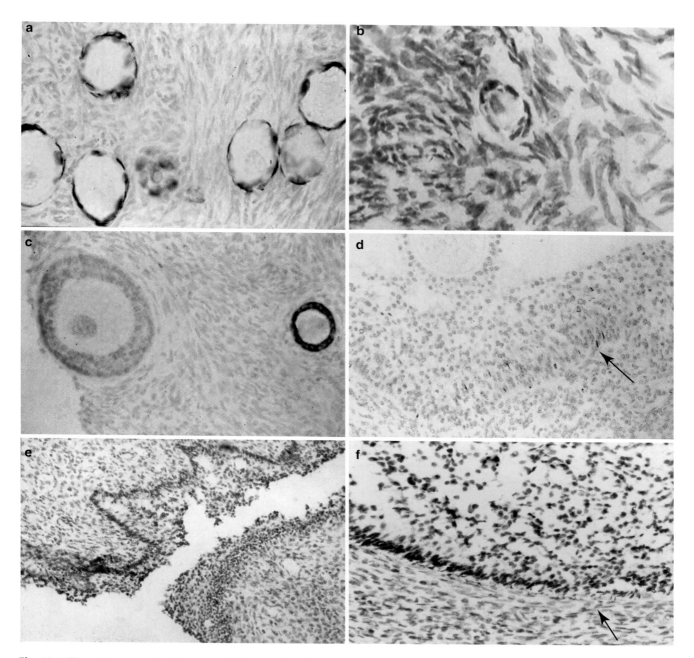

Fig. 10.2 The cyclic ovary of adult women depicts transient decrease of CK-positive follicle/granulosa cells. Immunohistology was performed with the pan-CK antibody Lu5. (**a–c**) CK expression decreases in primary follicles (compare **a** and **b**) and is missing in the preantral follicle (*left* in **c**). (**d**) The antral follicle develops CK-positive granulosa cells in the mural granulosa (*arrow*). (**e, f**) The preovulatory follicle with a gyrated wall contains many CK-positive granulosa cells. The basal layer of the mural granulosa consists of CK-positive cells in **f**. x a,b,c,d,f:240; e:130 (Adapted from Löffler et al. 2000)

10.3   Scheme of CK-Positive Cell Occurrence in Follicles Between Fetal Life and Ovulatory Period

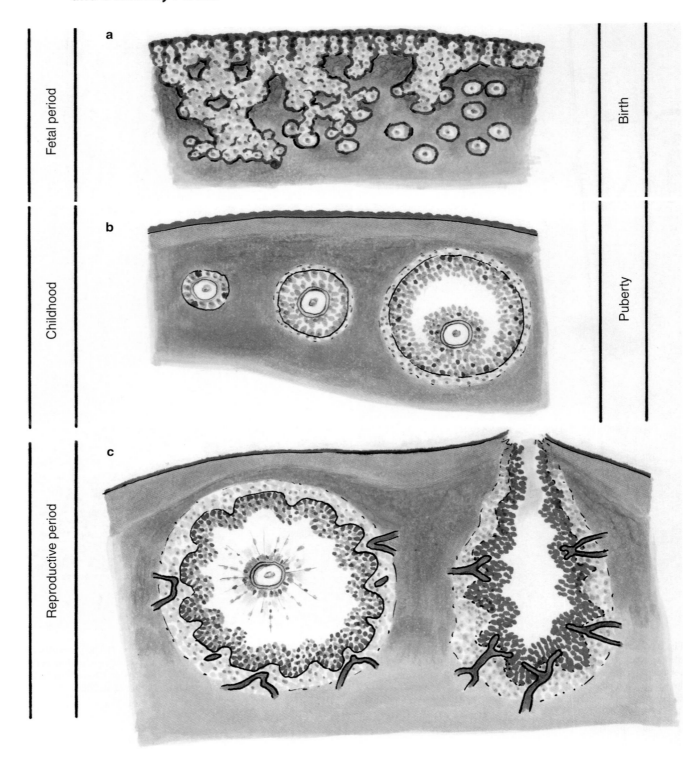

Fig. 10.3 The transient disappearance of CK-positive follicle cells is summarized. (**a**) In fetal life, the CK-positive surface epithelium develops CK-positive sex cords. Both structures are invaded by primordial germ cells. The medullary part of the sex cords segregates primordial follicles with the primary oocyte in the diplotene stage. All follicle cells are CK-positive cells. (**b**) Between birth and puberty, the preantral follicle lacks CK-positive cells. CK-positive cells reappear in the antral follicle. (**c**) In the ovulatory period of reproductive life, the number of CK-positive cells increases. It is at its maximum in the ruptured follicle (Drawn by R. Spanel)

10.4   Full Expression of CK-Positive Cells in the Developing Corpus Luteum

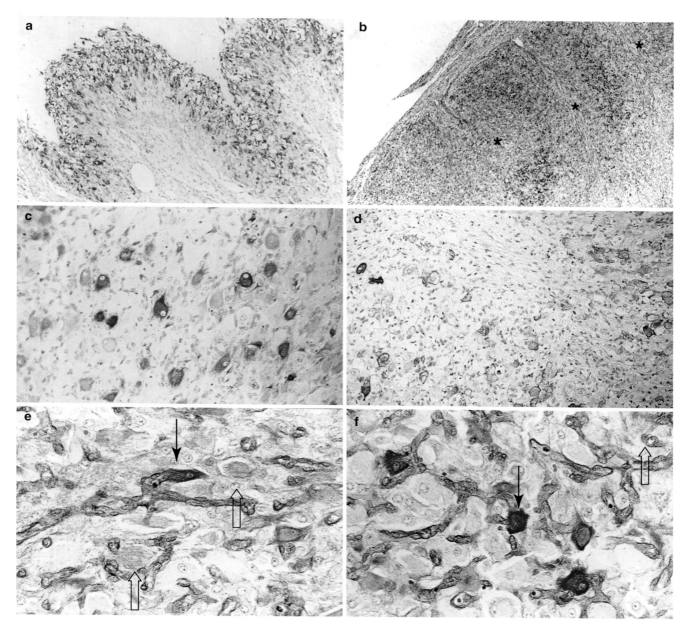

Fig. 10.4 The cyclic ovary of cows shows the highest number of CK-positive cells in the postovulatory period. Single immunostaining was performed with the pan-CK antibody Lu5 (**a–d**) and by double immunostaining for CK and for laminin (**e, f**). (**a**) The freshly ruptured follicle with a gyrated wall is rich in CK-positive granulosa cells. Most of the thecal cells are CK-negative. (**b**) In the developing CL, zonation occurs because of the infolded CK-positive granulosa cell layer and because of infoldings of the CK-negative thecal cell layer, which forms the septum (*asterisks*). (**c**) In the CL of the secretory stage, the intensity of CK expression varies between the CK-positive steroidogenic cells. Small and large CK-positive cells are uniformly spread in the parenchyma. (**d**) In the stage of regression, a few areas with weakly stained CK-positive cells remain. Their size defines them as steroidogenic cells. (**e, f**) The capillary bed in brown belongs to a CL in the secretory stage. The bed comprises single CK-positive vascular cells in dark red (*arrow*). They maintain a strong CK expression, which decreases in steroidogenic luteal cells (*open arrows*). x a.60; b:30; c,d:130; e,f:420 (Adapted from Ricken et al. 1995)

10.5   Different Phenotypes of Cell Cultures from the Corpus Luteum

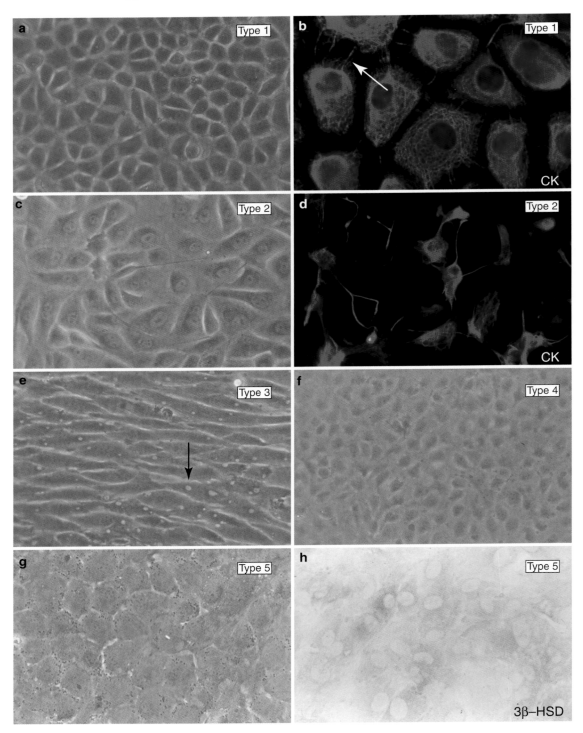

Fig. 10.5 Five different phenotypes of endothelial cells are isolated from the bovine CL and maintained in long-term culture. Type 1 represents a subtype of CK-positive endothelial cells. Type 2 correlates with a mixed culture. Types 3 and 4 are common endothelial cells. Type 5 might arise from steroidogenic luteal cells after loss of CK expression. Cultures in **a**, **c**, **e–g** are examined under the phase microscope, immunostained for CK in **b** and **d**, or studied by histochemistry for 3 β-hydroxysteroid dehydrogenase activity (3β-HSD) in **h**. (**a**, **b**) Type 1 produces a dense CK network with desmosome junctions (*arrow* in **b**). (**c**, **d**) Type 2 consists of CK-positive cells and of desmin-positive cells (not shown). (**e**, **f**) Type 3 shows spindle-shaped cells with a conspicuous intracellular vacuole (*arrow* in **e**), and type 4 consists of polygonal cells. (**g**, **h**) Type 5 depicts flat polygonal cells with lipid droplets and 3β-HSD activity. Types 1 and 5 are considered as dendritic-like cells with immunocompetence in luteolysis. x a,c-h:240; b:570 (Adapted from Spanel-Borowski 1991, 2011a, b, c; Spanel-Borowski and van der Bosch 1990; Fenyves et al. 1993; Spanel-Borowski et al. 1994a, b; Lehmann et al. 2000)

10.6   Scheme of CK-Positive Cell Occurrence in the Corpus Luteum and Concept of the Origin of Cell Culture Types 1 and 5

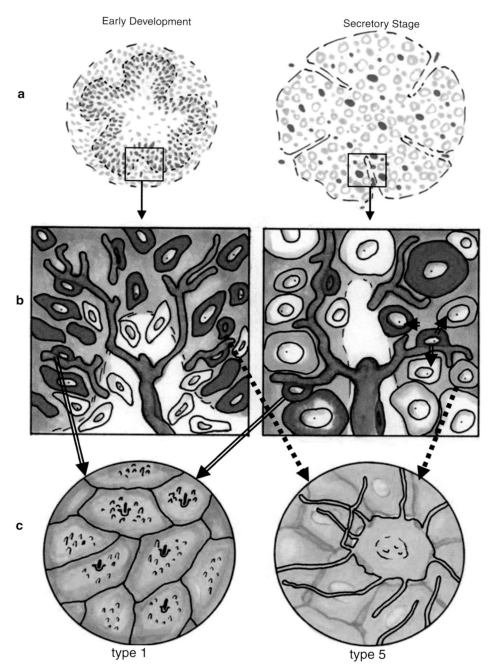

Fig. 10.6 The stage of development is seen on the *left*, and of secretion on the *right*. The *upper row* (**a**) represents a low magnification, the *middle row* (**b**) a high magnification, and the *lower row* (**c**) corresponds to the cell culture level. In early development, the former granulosa cell layer is infolded by the thecal cell layer. Most granulosa lutein cells are CK-positive. The former thecal cell layer, which forms the septum, rarely reveals a single CK-positive cell. The single CK-positive cell in the thecal layer is suggested to be an endothelial precursor cell, which subsequently associates with the microvascular bed, there maintaining CK expression (*green* in **b**). During the stage of secretion, steroidogenic luteal cells of small and large size are uniformly distributed in the parenchyma. The cells undergo a decrease and loss of CK expression indicated by *red, pink,* and *white* cells. The switching off of CK genes leads to the transition of steroidogenic CK-positive cells into granulosa-like cells of type 5 in culture (*broken arrow*). The microvascular CK-positive cells become type 1 cells in culture with maintenance of CK expression. They might sense danger and signal it to steroidogenic CK-positive cells at the onset of luteolysis (*short open arrows*, secretory stage). Types 1 and 5 are considered as dendritic-like cells because of characteristic functional responses (Drawn by R. Spanel. Adapted from Spanel-Borowski 2011a, b, c)

10.7   Albumin-Positive Cells in Follicles and Corpus Luteum

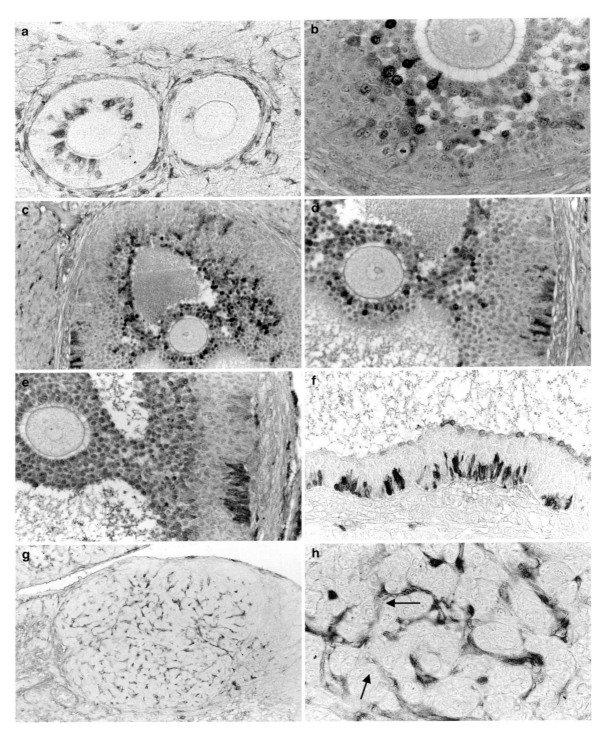

Fig. 10.7 Albumin-positive cells are detected in ovaries of golden hamsters (**a**, **f–h**) and rats (**b–e**) at proestrus (**a–f**) and early metestrus (**g**, **h**). Immunohistology was performed without H&E counterstaining (**a**, **f–h**) and with counterstaining (**b–e**). The small preantral follicle with negligible albumin-positive granulosa cells is considered to be intact. The adjacent follicle with strong albumin-positivity in the inner granulosa cells is likely starting atresia (Fig. 3.5c–e). (**b**) Albumin-positive cells are disseminated in the granulosa of a large preantral follicle. (**c**, **d**) The preovulatory follicle develops albumin-positivity in oocyte-associated cells, in antrum-lining cells, and in the basal layer of the mural granulosa cell layer. Albumin-positive cells in the basal layer compare with the similar occurrence of CK-positive granulosa cells in the preovulatory follicle in Fig. 10.2e, f. (**e**, **f**) The cumulus oophorus of the preovulatory follicle consists of albumin-positive cells and basal cells of columnar shape (**f**). (**g**, **h**) The CL in the secretory stage shows albumin-positive cells associated with the microvascular bed. The cell type is unclear because of nonresponsive cells (*arrows* in **f**). a:190; b:220; c:110; d,e,f:150, g:60; h:240 (Adapted from Spanel-Borowski 1987; Krishna and Spanel-Borowski 1989)

10.8 Up-regulation of Albumin-Positive Cells in IOR

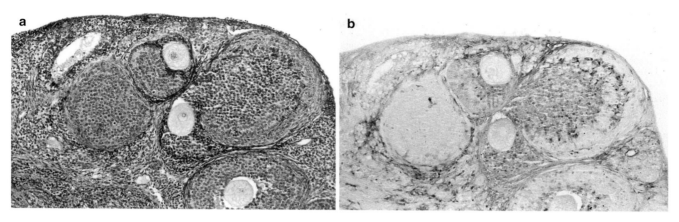

Fig. 10.8 There is a striking increase of albumin-positive granulosa cells in complete IOR from superovulated 29-day-old hamster ovaries on day 4 after PMSG application. The finding is deduced from serial sections immunostained for albumin without H&E counterstaining. (**a**) The herniated granulosa cells look intact. (**b**) Thus, albumin-positivity cannot be explained by cell death and unspecific background staining. Albumin-positive cells in IOR could compare with the increase in CK-positive granulosa cells in a freshly ruptured preovulatory follicle (Fig. 10.4a). x a,b:90

References

Antczak M, Van Blerkom J (2000) The vascular character of ovarian follicular granulosa cells: phenotypic and functional evidence for an endothelial-like cell population. Hum Reprod 15:2306–2318

Aust G, Brylla E, Lehmann I, Kiessling S, Spitzer K (1999) Different cytokine, adhesion molecule and prostaglandin receptor (PG-R) expression by cytokeratin 18 negative (CK−) and positive (CK+) endothelial cells (EC). Basic Res Cardiol 94:406

Ben-Ze'ev A, Amsterdam A (1989) Regulation of cytoskeletal protein organization and expression in human granulosa cells in response to gonadotropin treatment. Endocrinology 124:1033–1041

Czernobilsky B, Moll R, Levy R, Franke WW (1985) Co-expression of cytokeratin and vimentin filaments in mesothelial, granulosa and rete ovarii cells of the human ovary. Eur J Cell Biol 37:175–190

Davis EE, Brueckner M, Katsanis N (2006) The emerging complexity of the vertebrate cilium: new functional roles for an ancient organelle. Dev Cell 11:9–19

Elsadek B, Kratz F (2012) Impact of albumin on drug delivery – new applications on the horizon. J Control Release 157:4–28

Fenyves AM, Behrens J, Spanel-Borowski K (1993) Cultured microvascular endothelial cells (MVEC) differ in cytoskeleton, expression of cadherins and fibronectin matrix. A study under the influence of interferon-gamma. J Cell Sci 106(Pt 3):879–890

Fenyves AM, Saxer M, Spanel-Borowski K (1994) Bovine microvascular endothelial cells of separate morphology differ in growth and response to the action of interferon-gamma. Experientia 50:99–104

Knapp AC, Franke WW (1989) Spontaneous losses of control of cytokeratin gene expression in transformed, non-epithelial human cells occurring at different levels of regulation. Cell 59:67–79

Köhl J (2006) The role of complement in danger sensing and transmission. Immunol Res 34:157–176

Krishna A, Spanel-Borowski K (1989) Intracellular detection of albumin in the ovaries of golden hamsters by light and electron microscopy. Arch Histol Cytol 52:387–393

Lehmann I, Brylla E, Sittig D, Spanel-Borowski K, Aust G (2000) Microvascular endothelial cells differ in their basal and tumor necrosis factor-alpha-regulated expression of adhesion molecules and cytokines. J Vasc Res 37:408–416

Ley K, Gaehtgens P, Spanel-Borowski K (1992) Differential adhesion of granulocytes to five distinct phenotypes of cultured microvascular endothelial cells. Microvasc Res 43:119–133

Löffler S, Horn LC, Weber W, Spanel-Borowski K (2000) The transient disappearance of cytokeratin in human fetal and adult ovaries. Anat Embryol (Berl) 201:207–215

Merkwitz C, Ricken AM, Lösche A, Sakurai M, Spanel-Borowski K (2010) Progenitor cells harvested from bovine follicles become endothelial cells. Differentiation 79:203–210

Miller YI, Viriyakosol S, Binder CJ, Feramisco JR, Kirkland TN, Witztum JL (2003) Minimally modified LDL binds to CD14, induces macrophage spreading via TLR4/MD-2, and inhibits phagocytosis of apoptotic cells. J Biol Chem 278:1561–1568

Neumann E, Frei E, Funk D, Becker M, Schrenk H, Müller-Ladner U, Fiehn C (2010) Native albumin for targeted drug delivery. Expert Opin Drug Deliv 7:915–925

O'Neill LA, Bowie AG (2007) The family of five: TIR-domain-containing adaptors in Toll-like receptor signaling. Nat Rev Immunol 7:353–364

Ricken AM, Spanel-Borowski K, Saxer M, Huber PR (1995) Cytokeratin expression in bovine corpora lutea. Histochem Cell Biol 103:345–354

Ricken A, Rahner C, Landmann L, Spanel-Borowski S (1996) Bovine endothelial-like cells increase intercellular junctions under treatment with interferon-gamma. An in vitro study. Ann Anat 178:321–330

Santini D, Ceccarelli C, Mazzoleni G, Pasquinelli G, Jasonni VM, Martinelli GN (1993) Demonstration of cytokeratin intermediate filaments in oocytes of the developing and adult human ovary. Histochemistry 99:311–319

Serke H, Vilser C, Nowicki M, Hmeidan FA, Blumenauer V, Hummitzsch K, Lösche MT, Spanel-Borowski K (2009) Granulosa cell subtypes respond by autophagy or cell death to oxLDL-dependent activation of the oxidized lipoprotein receptor 1 and toll-like 4 receptor. Autophagy 5:991–1003

Serke H, Bausenwein J, Hirrlinger J, Nowicki M, Vilser C, Jogschies P, Hmeidan FA, Blumenauer V, Spanel-Borowski K (2010) Granulosa cell subtypes vary in response to oxidized low-density lipoprotein as regards specific lipoprotein receptors and antioxidant enzyme activity. J Clin Endocrinol Metab 95:3480–3490

Singla V, Reiter JF (2006) The primary cilium as the cell's antenna: signaling at a sensory organelle. Science 313:629–633

Spanel-Borowski K (1987) Immunocytochemical localization of albumin in ovarian follicles of fertile rats. Cell Tissue Res 248:699–702

Spanel-Borowski K (1991) Diversity of ultrastructure in different phenotypes of cultured microvessel endothelial cells isolated from bovine corpus luteum. Cell Tissue Res 266:37–49

Spanel-Borowski K (2011a) Footmarks of innate immunity in the ovary and cytokeratin-positive cells as potential dendritic cells, 209th edn, Advances in anatomy, embryology, and cell biology. Springer, Heidelberg. ISBN 978-3-642-16077-6

Spanel-Borowski K (2011b) Ovulation as danger signaling event of innate immunity. Mol Cell Endocrinol 333:1–7

Spanel-Borowski K (2011c) Five different phenotypes of endothelial cell cultures from the bovine corpus luteum: present outcome and role of potential dendritic cells in luteolysis. Mol Cell Endocrinol 338:38–45

Spanel-Borowski K, Bein G (1993) Different microvascular endothelial cell phenotypes exhibit different class I and II antigens under interferon-gamma. In Vitro Cell Dev Biol Anim 29A:601–602

Spanel-Borowski K, van der Bosch J (1990) Different phenotypes of cultured microvessel endothelial cells obtained from bovine corpus luteum. Study by light microscopy and by scanning electron microscopy (SEM). Cell Tissue Res 261:35–47

Spanel-Borowski K, Ricken AM, Kress A, Huber PR (1994a) Isolation of granulosa-like cells from the bovine secretory corpus luteum and their characterization in long-term culture. Anat Rec 239:269–279

Spanel-Borowski K, Ricken AM, Saxer M, Huber PR (1994b) Long-term co-culture of bovine granulosa cells with microvascular endothelial cells: effect on cell growth and cell death. Mol Cell Endocrinol 104:11–19

Stewart C, Stuart LM, Wilkinson K, van Gils JM, Deng J, Halle A, Rayner KJ, Boyer L, Zhong R, Frazier WA, Lacy-Hulbert A, Khoury JE, Golenbock DT, Moore KJ (2010) CD36 ligands promote sterile inflammation through assembly of a Toll-like receptor 4 and 6 heterodimer. Nat Immunol 11:155–161

Takeuchi O, Akira S (2010) Pattern recognition receptors and inflammation. Cell 140:805–820

Tscheudschilsuren G, Aust G, Nieber K, Schilling N, Spanel-Borowski K (2002) Microvascular endothelial cells differ in basal and hypoxia-regulated expression of angiogenic factors and their receptors. Microvasc Res 63:243–251

Tsikolia N, Merkwitz C, Sass K, Sakurai M, Spanel-Borowski K, Ricken A (2009) Characterization of bovine fetal Leydig cells by KIT expression. Histochem Cell Biol 1322:623–632

Wolf KW, Spanel-Borowski K (1992) The interphase microtubule cytoskeleton of five different phenotypes of microvessel endothelial cell cultures derived from bovine corpus luteum. Tissue Cell 24:347–354

Aspects of Immunological Control, Novel Concepts, Challenges, and Clinical Perspectives

11

11.1 Immunological Control and Novel Concepts

Ovarian tissues in transformation and in decay are potential places of danger, which puts cells under stress. The subsequent misbalance in the oxidative system in favor of higher oxidant than antioxidant levels leads to the production of ROS. Mild ROS production appears to stimulate cell processes such as proliferation, differentiation, and angiogenesis (Dandapat et al. 2007). High ROS levels harm cells severely, which release danger signals that are recognized by dendritic-like cells in the ovary (Kohchi et al. 2009). This calls up the physiological side of INIM function that is substantially involved in the plasticity of the ovary through pro- and anti-inflammatory actions. Thus, ovarian places of tissue turnover can be classified into nondangerous and dangerous sites. Follicular development with ingrowth from the cortical periphery toward the medulla remains nonresponsive in terms of cell death and inflammatory response. Atresia of preantral follicles through pathway A also keeps quiescent any tissue response apart from the transition of follicle cells into the cortical stroma and the resumption of meiosis (Fig. 3.2). The constructive effects might be explained by low ROS levels. In contrast, the strong inflammatory response in antral follicles under atresia pathway B indicates a dangerous site with rather high ROS levels (Fig. 3.3). Tissue danger asks for INIM protection. The granulosa cell subtypes such as dark cells, albumin-positive cells, and CK-positive cells, which augment in regressing follicles (Fig. 3.5), might represent the immunological defense arm for tissue repair and restitution of function.

The interstitial cortical tissue appears to be a special place of immunological defense that was previously overlooked. The ovarian cortex is densely populated by CD18-positive leukocytes containing a high percentage of mast cells (Figs. 4.1c–e and 9.2a, e). The "cortical" leukocytes together with mast cells must have a function of their own, because follicles and CLs recruit leukocytes without mast cells. Cellular stromatolysis of mature interstitial gland cells is considered a novel form of inactivation/degeneration, being associated with shedding of cytoplasmic components into the cortical stroma. The event could be involved in up-regulating endothelial cell adhesion molecules and thus account for the CD18-positive cell pool with mast cells. The presence of mast cells might be beneficial for the control of INIM and adaptive immunity responses (Kalesnikoff and Galli 2008; Shelburne and Abraham 2011). Of note, mast cells are undetectable in the cortical rim and in the postmenopausal ovary lacking mature interstitial gland cells (Fig. 9.2).

The mature follicle before and after rupture compares with a battlefield. Dangerous oxidative stress finally causes the ovulatory event, according to the innovative concept of INIM control (Spanel-Borowski 2011a, b). The ovulatory process starts as an inside-out process initiated by oxidative stress. It is caused by high ROS production either due to full-speed steroidogenesis or to oxLDL increase that triggers the LOX-1-dependent cascade with production of ROS as byproducts. Apoptotic death ensues in CK-negative granulosa cells with the release of alarmins as ligands of INIM. Alarmins and oxLDL recognize CK-positive granulosa cells expressing TLR4. Receptor activation leads to the proinflammatory Myd88-dependent pathway with nonapoptotic cell death, angiogenesis, recruitment of leukocytes, and connective tissue degradation (O'Neill and Bowie 2007; Takeuchi and Akira 2010). The subsequent outside-in process from the thecal headquarter toward the granulosa cell layer is judged as a repair event related to the development of the CL. It requires the help of KIT-positive thecal cells, which intermingle with CK-positive granulosa cells. Luteinization is mediated together with maturation of the microvascular bed and connective tissue ingrowth into the former antrum. The outside-in event seems to use the anti-inflammatory TRIF-dependent pathway of TLR4 signaling. INIM is seen to be a successful manager also in superovulated ovaries with many IORs as multiple dangerous sites (Figs. 7.7, 7.8, 7.9, and 7.10). The tremendous tissue damage appears to heal quickly because ovarian cycles are resumed on time.

K. Spanel-Borowski, *Atlas of the Mammalian Ovary*,
DOI 10.1007/978-3-642-30535-1_11, © Springer-Verlag Berlin Heidelberg 2012

Structural luteolysis in species with a long ovarian cycle like the cow represents a chronic inflammatory event that seems to require the alliance of adaptive immunity (Spanel-Borowski 2011c). Type 1 cells with CK filaments and type 5 with long filipodia are suggested to be cells with immunocompetence. Their interaction as a luteovascular unit in control of luteolysis is depicted in Fig. 11.1. The microvascular type 1 CK-positive endothelial cell senses danger such as hypoxia in the lumen with the help of the single nonmotile cilium recognized as a chemosensory antenna with proteins of signal transduction (Davis et al. 2006; Singla and Reiter 2006). Two different pathways could then be initiated from microvascular type 1 cells. One pathway is responsible for the recruitment of monocytes and T cells, whereas the other pathway generates IFN-γ. The cytokine then causes the steroidogenic luteal subtype to lose CK filaments, long cell processes are developed, and MHC II is up-regulated in type 5 cells. Cell processes become the docking station for naïve T cells transforming into cytotoxic and helper cells (Takeda and Akira 2005). In addition, TNF-α secretion from type 5 cells leads to the immediate death of common steroidogenic luteal cells. This novel concept of luteolysis remains provocative and speculative as long as the true existence of dendritic-like cells in follicles and CLs awaits confirmation. Collectively, INIM actions in the ovary are deduced from inflammatory and anti-inflammatory events that do not harm tissue integrity. In the near future, novel insights into INIM-specific pathways in immunocompetent cells of the ovary will reveal the precise molecular patterns closely linked to organ-specific immunity (Matzinger 2007; Matzinger and Kamala 2011).

Appealing figure sequences are provided about tissue disintegration beyond the common knowledge of necrotic/apoptotic cell death. Presently, survival autophagy is considered as the only possibility of removing unwanted organelles after cell damage (Klionsky 2007; Chen and Klionsky 2011). Cytoplasmic shedding seen here in mature interstitial gland cells (Fig. 4.3) might be a new aspect of cell degeneration. Cytoplasmic shedding might also be conducted by other endocrine cells in the case of shortage in stimulation. Uncontrolled signaling of cytoplasmic shedding might lead to necrotic cell death as is known for cell-death autophagy. Bulk tissue removal through herniation of cells into damaged microvessels termed vascular stromatolysis or vascular luteolysis provides another unusual view on rapid tissue removal (Figs. 8.1, 8.2, 8.3, and 8.4). This exciting possibility represents an elegant device with which to immediately maintain homeostasis without additional tissue lesion. Theoretically, fine modulation of INIM signaling has the capacity to selectively degrade vessel walls for bulk tissue removal. It is only INIM with its wide spectrum of diverse signaling pathways that can mediate such an incredible occurrence. A malignant tumor might spontaneously heal by bulk tissue removal similar to vascular luteolysis.

11.2 Challenges and Clinical Perspectives

Big challenges lie ahead in understanding the immunoregulatory pathways in the ovary at the molecular and cellular level. Research on this area becomes more complex by integration of the endocrine system, presently the established master of ovarian functions. Most importantly, the concept of CK-positive cells as dendritic-like cells needs final proof by in situ localization of dendritic antigens in follicles and in CLs (Banchereau and Steinman 1998; Steinman and Banchereau 2007). In vitro studies on signaling pathways and immunoresponses are feasible, because the candidates of potential dendritic cells are isolated from human follicles and from bovine CLs. Moreover, the cellular criteria are well characterized (Spanel-Borowski 2011a, c). In vivo research can be performed in rodents with synchronized ovarian cycles after gonadotropic stimulations. The procedure allows an exact time sequence to be obtained for the activation of cytokines such as IFN-γ, TNF-α, and TGF-β in superovulated ovaries. The cytokine maxima might differ in golden hamsters and in white-footed mice undergoing structural luteolysis or vascular stromatolysis, respectively. In comparison, cytokine levels could be altered in polycystic ovaries of DHEA-treated mice. Golden hamster ovaries, in which mature interstitial gland cells are replaced by inactive interstitial gland cells after a change in housing from long to short photoperiods, are a nice model for studying cytoplasmic shedding. Additionally, ovaries of white-footed mice represent a physiological model for assessing vascular luteolysis. Gene-manipulated mice with knocked out genes for TLR receptors or with green fluorescent-labeled promoter in control of a dendritic-cell-specific antigen will contribute toward our understanding of INIM functions in the ovary.

All efforts to understand the immunoregulation in the ovary provide far-reaching clinical perspectives. Hypoactivation of INIM signaling might be responsible for anovulation disorders like polycystic ovary syndrome. For patients who show resistance to pharmacological treatment, ovarian drilling has proved to be successful (Api 2009). The success can be explained by "danger application," which appears to awaken the ovarian INIM function. A disturbed INIM cascade can be the cause for CL insufficiency or for the unruptured and luteinized follicle syndrome (Qublan et al. 2006). Progesterone serum rises, but no oocyte is released into the fallopian tube and no fertilization is recorded. The syndrome might hide the IOR event and cause a "false" CL in women under IVF therapy. A decrease of INIM function in the postmenopausal ovary because of ovarian cycle cessation could contribute to the development of epithelial ovarian cancer. It is a tumor of senior age with low hope for the patient's recovery (Kurman and Shih 2010). Therapeutic strategies could be improved by analyzing molecules of the potential dendritic-like CK-positive cells. Residences of primordial and primary follicles are suggested to be nondangerous places. Hyperactivation of the INIM cascade might be the cause for hyperstimulation syndrome with life-threatening release of VEGF (Kahnberg et al. 2009). Increase in INIM function could also lead to premature ovarian failure and autoimmune oophoritis (Vujovic 2009). The examples of ovarian disorders postulated to be regulated through hypo- and hyperactivation of INIM signaling are presumably initiated by oxidative stress of different intensities. It is conceivable that daily treatment with the antioxidant melatonin brings back regular ovarian cycles. The success of the therapy might depend on varying degrees of oxidative stress, thus explaining conflicting results of melatonin treatment in women (Srinivasan et al. 2009).

Progress in science compares with helically aligned floors in a flight of stairs. Lower floors belong to the past and higher floors to the future. By reaching higher floors each domain is extended, modified, and changed. Understanding the dynamics in the ovary compares with spirals of a never-ending story (Fig. 11.2). This atlas intends to shed light on the complexity of ovarian dynamics and to demonstrate that morphology remains one of the gold standards in the exciting field of INIM control in the ovary.

11.3   Scheme for CK-Positive Cells as a Danger Sensor in the Microvascular Bed: Novel Concept of a Luteovascular Unit

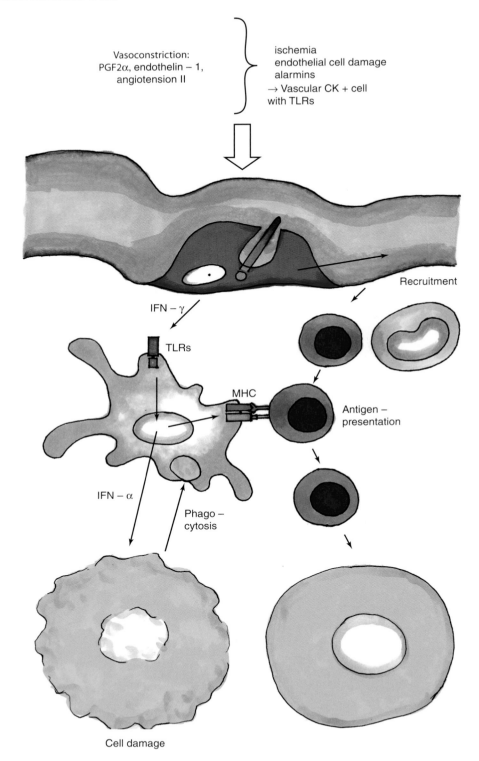

Fig. 11.1 The concept of early structural luteolysis depends on the luteovascular unit. Danger signals from ischemia-damaged normal endothelial cells are sensed by microvascular CK-positive cells with TLRs. The signaling cascade generates the gateways either for leukocyte recruitment, in particular T cells, or for IFN-γ secretion. The cytokine affects steroidogenic cells with TLRs, which have been converted from steroidogenic CK-positive cells by losing CK expression and developing long cell processes. Interferon-γ activates the TNF-α gene, and the TNF-α protein harms normal steroidogenic cells. Interferon-γ also turns the converted cells into antigen-presenting cells by up-regulation of the MHCII complex for the presentation of the antigen from damaged cells. Naive T cells are trained into T cell subsets, which mediate different cell death forms. *Arrows* represent theoretical associations (Drawn by R. Spanel; adapted from Spanel-Borowski 2011c)

11.4   Spiral as a Symbol of the Ovary

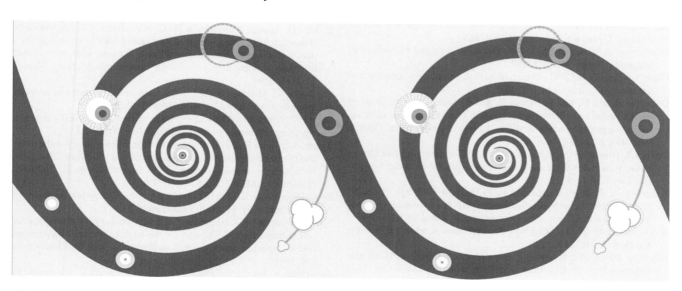

Fig. 11.2 In the reproductive period, ovarian cycles compare with dynamic spirals. Immature small follicles arise in the center from where the cohort of antral follicles is recruited to provide the dominant follicle, which becomes the preovulatory/Graaf follicle. After expulsion of the mature oocyte, the follicle wall turns into a CL. The next cycle appears on the horizon (Designed by R. Spanel)

References

Api M (2009) Is ovarian reserve diminished after laparoscopic ovarian drilling? Gynecol Endocrinol 25:159–165

Banchereau J, Steinman RM (1998) Dendritic cells and the control of immunity. Nature 392:245–252

Chen Y, Klionsky D (2011) The regulation of autophagy – unanswered questions. J Cell Sci 124:161–170

Dandapat A, Hu C, Sun L, Mehta JL (2007) Small concentrations of oxLDL induce capillary tube formation from endothelial cells via LOX-1-dependent redox-sensitive pathway. Arterioscler Thromb Vasc Biol 27:2435–2442

Davis EE, Brueckner M, Katsanis N (2006) The emerging complexity of the vertebrate cilium: new functional roles for an ancient organelle. Dev Cell 11:9–19

Kahnberg A, Enskog A, Brännström M, Lundin K, Bergh C (2009) Prediction of ovarian hyperstimulation syndrome in women undergoing in vitro fertilization. Acta Obstet Gynecol Scand 88:1373–1381

Kalesnikoff J, Galli SJ (2008) New developments in mast cell biology. Nat Immunol 9:1215–1223

Klionsky DJ (2007) Autophagy: from phenomenology to molecular understanding in less than a decade. Nat Rev Mol Cell Biol 8:931–937

Kohchi C, Inagawa H, Nishizawa T, Soma G (2009) ROS and innate immunity. Anticancer Res 29:817–821

Kurman RJ, Shih I (2010) The origin and pathogenesis of epithelial ovarian cancer: a proposed unifying theory. Am J Surg Pathol 34:433–443

Matzinger P (2007) Friendly and dangerous signals: is the tissue in control? Nat Immunol 8:11–13

Matzinger P, Kamala T (2011) Tissue-based class control: the other side of tolerance. Nat Rev Immunol 11:221–230

O'Neill LA, Bowie AG (2007) The family of five: TIR-domain-containing adaptors in Toll-like receptor signaling. Nat Rev Immunol 7:353–364

Qublan H, Amarin Z, Nawasreh M, Diab F, Malkawi S, Al-Ahmad N, Balawneh M (2006) Luteinized unruptured follicle syndrome: incidence and recurrence rate in infertile women with unexplained infertility undergoing intrauterine insemination. Hum Reprod 21:2110–2113

Shelburne C, Abraham S (2011) The mast cell in innate and adaptive immunity. Adv Exp Med Biol 716:162–185

Singla V, Reiter JF (2006) The primary cilium as the cell's antenna: signaling at a sensory organelle. Science 313:629–633

Spanel-Borowski K (2011a) Footmarks of innate immunity in the ovary and cytokeratin-positive cells as potential dendritic cells, 209th edn, Advances in anatomy, embryology, and cell biology. Springer, Heidelberg. ISBN 978-3-642-16077-6

Spanel-Borowski K (2011b) Ovulation as danger signaling event of innate immunity. Mol Cell Endocrinol 333:1–7

Spanel-Borowski K (2011c) Five different phenotypes of endothelial cell cultures from the bovine corpus luteum: present outcome and role of potential dendritic cells in luteolysis. Mol Cell Endocrinol 338:38–45

Srinivasan V, Spence W, Pandi-Perumal S, Zakharia R, Bhatnagar K, Brzezinski A (2009) Melatonin and human reproduction: shedding light on the darkness hormone. Gynecol Endocrinol 25:779–785

Steinman RM, Banchereau J (2007) Taking dendritic cells into medicine. Nature 449:419–426

Takeda K, Akira S (2005) Toll-like receptors in innate immunity. Int Immunol 17:1–14

Takeuchi O, Akira S (2010) Pattern recognition receptors and inflammation. Cell 140:805–820

Vujovic S (2009) Aetiology of premature ovarian failure. Menopause Int 15:72–75

Index

K. Spanel-Borowski, *Atlas of the Mammalian Ovary*,
DOI 10.1007/978-3-642-30535-1, © Springer-Verlag Berlin Heidelberg 2012

Printing: Ten Brink, Meppel, The Netherlands
Binding: Stürtz, Würzburg, Germany